○ ○ ● ○

Mathematics
in Daily Living
REVISED

MEASUREMENT
and GEOMETRY

Nerissa Bell Bryant
Staff Consultant
Adult Performance Level Project
Louisiana Tech University

Loy Hedgepeth
Director of Adult Education
Ouachita Parish Schools
Monroe, Louisiana

Steck-Vaughn Company ○ **Austin, Texas**

Introduction

This book contains adult-oriented instructional material designed to teach mathematics skills and life-coping skills to the mature learner.

The academic skill content of this book was determined by data compiled from a survey of adult education teachers. This survey revealed the topics for which teachers most urgently needed teaching materials.

In addition to the mathematics skill content, this book focuses on life-coping skills needed daily by adults. The curriculum emphasis focuses on these five general areas: health, government and law, consumer economics, community resources, and occupational knowledge. Information on various topics in these five areas is presented along with practice of mathematics skills. Thus, as learners progress through the mathematics units, they are afforded an opportunity to (1) develop academic skills related to *mathematics competence* and (2) develop life-coping skills related to *functional competence*.

During their development these materials were field-tested at the Northeast Louisiana Learning Center in Monroe, Louisiana.

How To Use This Book

Each unit contains individualized instruction for self-development. However, the format and use of the book should be understood in order to expedite progress and maximize proper use of the material.

Unit:	A section providing instruction, examples, practices, reviews, and evaluations of a designated skill. Units are subdivided into lessons.
Skill:	Each intermediate skill necessary to the mastery of the designated unit skill is presented in each lesson title.
Instruction:	Explanation and/or definition of each skill is offered. Examples follow each segment of instruction to clarify the instruction and prepare the learner to undertake practice exercises.
Exercises:	Exercises to be worked and checked by the learner. Exercise B provides the same type of practice as Exercise A.
Review:	An exercise providing practice and review on skills taught in the unit.
Coping Skills:	Most units end with an activity designed to develop greater competency in life-coping skills. Other activities related to coping skills are within the lessons.
Answers:	Answers to all pretests, exercises, and reviews are provided at the back of the book.

NOTICE: Answer Key is bound in the back of the book.

ISBN 0-8114-1515-5

4 5 6 7 8 9 0—89 88

Contents

UNIT 1—PRETEST
MEASUREMENT

Add, subtract, multiply, or divide as indicated. Do not show fractions in your answers.

1. 14 ft. 8 in.
+ 3 ft. 8 in.

2. 2 qt. 1 pt.
× 4

3. 7 gal. 2 qt.
− 3 gal. 3 qt.

4. 8 pk. 4 qt. ÷ 4 =

5. 5 lb. 5 oz.
+ 6 lb. 15 oz.

6. 9 yd. 1 ft.
− 5 yd. 2 ft.

Circle the smaller unit in each group.

7. 1 centimeter or 1 millimeter

8. 1 dekagram or 9 grams

9. 50 kiloliters or 50 milliliters

Fill in the blanks.

10. The basic metric unit for measuring mass is a _____.

11. What is the basic metric unit for measuring capacity? _____

12. How many degrees are on the Celsius thermometer between the freezing and boiling points of water? _____

Convert these measures.

13. 4 ounces ≈ _____ grams

14. 2 pounds 2 ounces ≈ _____ grams

15. 56 grams ≈ _____ ounces

16. 1.1 pounds ≈ _____ kilogram

Answer the following questions.

17. For which temperature will you need heavy clothing, 27°F or 20°C? _____

18. What is normal body temperature on the Celsius scale? _____

19. Which is warmer, 70°F or 70°C? _____

20. Would a person be comfortable bathing in 98.6°C water? _____

1

MEASUREMENT

LESSON ONE: Understanding and Using English Units of Weights and Measures

Instruction

The table below lists some of the **English** units of weights and measures.

English Weights and Measures

Length
12 inches (in.) = 1 foot (ft.)
3 feet = 1 yard (yd.)
5.5 yards = 1 rod (rd.)
5,280 feet = 1 mile (mi.)
320 rods = 1 mile

Weight
16 ounces (oz.) = 1 pound (lb.)
2,000 pounds = 1 ton

Liquid Measure
8 fluidounces (fl. oz.) = 1 cup (c.)
2 cups = 1 pint (pt.)
2 pints = 1 quart (qt.)
4 quarts = 1 gallon (gal.)

Dry Measure
2 pints = 1 quart
8 quarts = 1 peck (pk.)
4 pecks = 1 bushel (bu.)

To express a given amount in smaller units, multiply the amount by the number of smaller units contained in one larger unit.

Example

Change 5 pounds to ounces.

1 pound = 16 ounces
16 × 5 = 80

5 pounds = 80 ounces

To express a given amount in larger units, divide the amount by the number of smaller units contained in one larger unit.

Example

Change 72 inches to feet.

1 foot = 12 inches
72 ÷ 12 = 6

72 inches = 6 feet

Adding and subtracting weights and measures in which different units are used together is similar to adding and subtracting fractions.

2

Example

> 5 ft.   8 in.
> + 7 ft.   6 in.
> 12 ft. 14 in. = 13 ft. 2 in.
>
> Since 14 inches is more than 1 foot but less than 2 feet, one unit (1 foot) is added to the larger unit of measure.

Example

> 4 gal. 1 qt. = 3 gal. 5 qt.
> − 1 gal. 2 qt. = 1 gal. 2 qt.
>   2 gal. 3 qt.
>
> Since 2 quarts cannot be subtracted from 1 quart, 4 quarts (1 gal.) was borrowed from 4 gallons and added to 1 quart before subtracting.

Measures can also be multiplied and divided.

Example

> 8 lb. 10 oz.
>   × 5 oz.
> 40 lb. 50 oz. = 43 lb. 2 oz.

Example

> 12 ft. 8 in. ÷ 8
>
> 12 ft. ÷ 8 = 1 ft. with 4 feet left over. 4 feet = 48 inches. Add this to the 8 inches in the problem and divide by 8 again.
>
> 8 in. + 48 in. = 56 in.
> 56 in. ÷ 8 = 7 in.
>
> The answer is 1 foot 7 inches.

Sometimes, it is easier to write the quantities as fractions before adding, subtracting, multiplying, or dividing.

Example

> 5 gal. 3 qt. = $5\frac{3}{4}$ gal.
> + 3 gal. 2 qt. = $3\frac{2}{4}$ gal.
> $8\frac{5}{4}$ gal. = $9\frac{1}{4}$ gal.
>
> $9\frac{1}{4}$ gal. can be written as 9 gal. 1 qt. if you need the answer in that form.

Exercise A

Add, subtract, multiply, or divide as indicated. Do not show fractions in your answers.

1. 12 ft. 10 in.
    + 5 ft. 4 in.

2. 6 qt. 1 pt.
    × 4

3. 8 gal. 2 qt.
    − 5 gal. 3 qt.

4. 10 pk. 4 qt. ÷ 7 =

5. 2 lb. 12 oz.
    + 3 lb. 8 oz.

6. 5 yd. 1 ft.
    − 2 yd. 2 ft.

7. 8 mi. 160 rd.
    × 2

8. 5 c. 7 fl. oz.
    − 2 c. 5 fl. oz.

9. 8 yd. 1 ft. ÷ 5 =

10. 13 ft. 7 in.
    + 2 ft. 9 in.

11. 6 lb. 4 oz.
    − 4 lb. 10 oz.

12. 7 yd. 2 ft.
    × 5

13. 6 gal. 2 qt.
    − 4 gal. 3 qt.

14. 4 ft. 11 in.
    × 5

15. 3 lb. 4 oz.
    − 1 lb. 8 oz.

Exercise B

Add, subtract, multiply, or divide as indicated. Do not show fractions in your answers.

1. 3 yd. 1 ft.
    × 9

2. 6 gal. 1 qt.
    − 4 gal. 3 qt.

3. 4 mi. 180 rd.
    + 8 mi. 140 rd.

4. 18 lb. 12 oz. ÷ 4 =

5. 6 bu. 3 pk.
    × 6

6. 10 gal. 2 qt.
    + 6 gal. 3 qt.

7. 7 yd. 2 ft.
    − 7 yd. 1 ft.

8. 4 qt. 1 pt. ÷ 3 =

9. 3 ft. 9 in.
    × 5

10. 3 lb. 6 oz.
    × 5

11. 4 gal. 3 qt.
    + 1 gal. 1 qt.

12. 18 ft. 9 in.
    − 5 ft. 5 in.

Understanding and Using Metric Units to Measure Length

Instruction

The **metric system** of measure is based upon units of ten. Most of the major countries of the world use the metric system. In the mid-1970s, the United States began a gradual changeover to the metric system.

The basic unit of length in the metric system is the **meter** (m). Notice that the symbol for meter is written without a period, as are all metric symbols. A meter is a little longer than a yard.

Units of length larger and smaller than a meter are formed by adding prefixes to the basic unit, meter. The table below shows the prefixes and their meanings.

Prefix	Meaning	Example
kilo-	1,000 times basic unit	1 kilometer (km) = 1,000 meters
hecto-	100 times basic unit	1 hectometer (hm) = 100 meters
deka-	10 times basic unit	1 dekameter (dam) = 10 meters
deci-	.1 times basic unit	1 decimeter (dm) = .1 meter
centi-	.01 times basic unit	1 centimeter (cm) = .01 meter
milli-	.001 times basic unit	1 millimeter (mm) = .001 meter

Metric measures are usually written as **decimals**. This makes changing from one unit to another one easy. All you need to do is move the decimal point to the left or the right.

This staircase shows that we need to move the decimal point to the **right** to go from one unit to a smaller one and to the **left** to go from one unit to a larger one. Move the decimal point one place for each step.

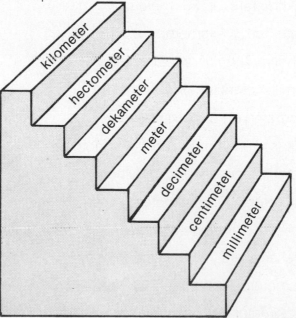

Example

Change 26.8 meters to centimeters.

26.8 meters ⟶ 26.80. ⟶ 2,680 centimeters

Example

Change 50.2 decimeters to kilometers.

50.2 decimeters ⟶ .0050.2 ⟶ .00502 kilometers

When comparing unlike units, it is sometimes necessary to convert to like units in order to decide which is larger or smaller.

Example

Which is larger, 48 millimeters or 23 decimeters?

To solve, either change 48 millimeters to decimeters, or change 23 decimeters to millimeters.

48 millimeters = .48. decimeters

23 decimeters is larger than 48 millimeters because 23 is greater than .48.

Exercise A

Decide which unit in each group is smallest and circle it.

1. 1 meter or 1 decimeter

2. 1 hectometer or 1 dekameter

3. 1 centimeter or 1 millimeter

4. 1 meter or 1 centimeter

5. 6 dekameters or 83 meters

6. 1 meter or 1 kilometer

7. 50 centimeters or 2 decimeters

8. 200 millimeters or 2 centimeters

9. 1 kilometer or 5 centimeters

10. 1 dekameter, 1 millimeter, 1 hectometer

11. 1 meter, 1 centimeter, 1 kilometer

12. 2 kilometers, 2 centimeters, 1 decimeter, 1 meter

13. 5 meters, 1 meter, 10 centimeters, 2 kilometers

14. 23 dekameters, 1 kilometer, 56 meters, 1,000 millimeters

15. 2 kilometers, 1 meter, 800 millimeters

16. 1 hectometer, 10 dekameters, 2,000 millimeters

6

17. 86 meters, 1 kilometer, 20 hectometers

18. 10 decimeters, 100 centimeters, 200 millimeters

19. 1 dekameter, 10 hectometers, 100 millimeters

20. 1 meter, 10 meters, 100 meters, 1,000 meters

Exercise B   Circle the largest unit in each group.

1. 1 centimeter or 1 decimeter

2. 1 meter or 2 millimeters

3. 1 hectometer or 150 meters

4. 3 kilometers or 2,000 centimeters

5. 586 millimeters, 1 meter, 275 centimeters

6. 10 dekameters, 2 hectometers, 1 kilometer

7. 3 kilometers, 1 meter, 100 millimeters

8. 55 centimeters or 7 centimeters

9. 15 centimeters or 10 kilometers

10. 2 centimeters or 2 kilometers

11. 1 kilometer or 999 meters

12. 1 millimeter, 1 kilometer, 1 meter

13. 3 meters, .1 meter, 0.001 meter

14. 9 dekameters or 91 meters

15. 47 meters or 50 dekameters

16. 2 kilometers, 4 meters, 6 centimeters

17. 1 decimeter or 1 meter

18. 1 meter, 9 decimeters, 200 millimeters

19. 1 centimeter or 1 millimeter

20. 1 decimeter, 1 centimeter, 1 meter

LESSON THREE: Comparing Metric and English Units of Length

Instruction

The most common units of measure in the English system are the **inch**, **foot**, **yard**, and **mile**. We shall compare these to units in the metric system. You will notice that we use a sign (≈) which means "approximately equal to" since they are not equal.

Example

> 1 centimeter ≈ .4 inch
> 1 meter ≈ 3.3 feet
> 1 meter ≈ 1.1 yards
> 1 kilometer ≈ .6 mile

Exercise A

Decide what unit of measure you would probably use to measure the things below. Give both English and metric units. Choose from these: inch, foot, yard, mile, centimeter, meter, and kilometer.

1. height of a door

2. length of a pencil

3. distance between two cities

4. width of a fingernail

5. length of a swimming pool

6. length of a pocket comb

7. height of an adult

8. width of a refrigerator

9. length of a car

10. length of a zipper

11. width of a window

12. length of a match

13. width of a child's foot

14. width of a drawer

15. width of a room

16. height of a chimney

17. length of a bed

18. width of a toothbrush

8

Courtesy Plaskolite, Inc.

Exercise B

Give the English and metric units of measure you would probably use to measure the following things. Choose from these: inch, foot, yard, mile, centimeter, meter, and kilometer.

1. width of a book

2. a person's waist

3. a postage stamp's width

4. the diagonal of a TV screen

5. a city block

6. length of a patio

7. length of a paper clip

8. height of a TV antenna

9. width of an envelope

10. height of a wall

11. height of a tall tree

12. length of a boat

13. length of a rope

14. height of an infant

15. depth of a fireplace

16. length of a screw

17. height of a chair

18. a lake's depth

19. height of a box

20. width of a bulletin board

LESSON FOUR: Understanding the Relationship Between Millimeters and Centimeters

Instruction

A **centimeter** (cm) is divided into 10 equal parts. Each part is called a **millimeter** (mm). A millimeter is $\frac{1}{1000}$, or 0.001, of a meter.

Example

Exercise A

Using a ruler divided into centimeters and millimeters, measure the length of each object to the nearest centimeter.

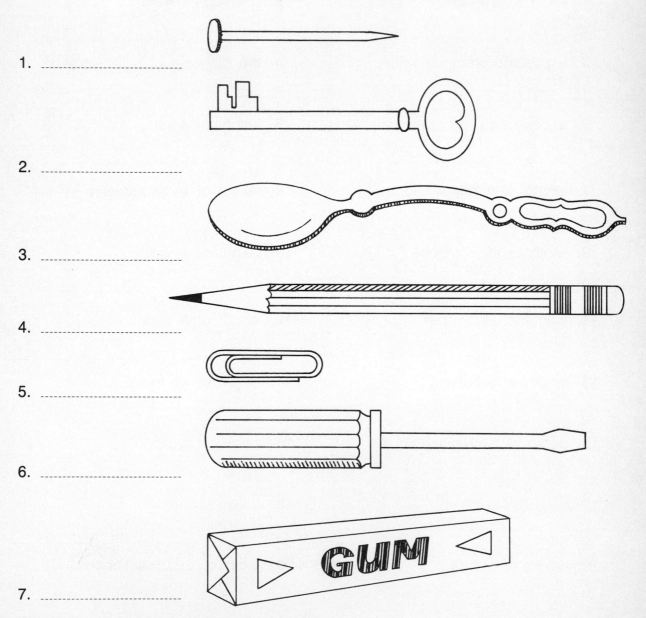

1. _____

2. _____

3. _____

4. _____

5. _____

6. _____

7. _____

Exercise B

Measure the length of each object to the nearest centimeter.

1. ....

2. ....

3. ....

4. ....

5. ....

6. ....

7. ....

8. ....

CANDY

CRAYON

Using Metersticks

Instruction

A **meter** (m) is divided into 10 equal parts. Each part is called a **decimeter** (dm). Each decimeter is divided into 10 **centimeters**. Each centimeter is divided into 10 **millimeters**.

Example

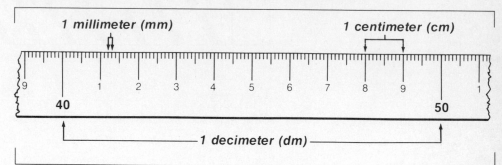

Exercise A

Use a meterstick to measure the following things.

1. length of your room _____

2. height of a window _____

3. width of a book _____

4. height of a friend _____

5. length of a car _____

6. length of a belt _____

7. width of a table _____

Exercise B

Use a meterstick to measure the following things.

1. width of a refrigerator _____

2. length of a bed _____

3. height of a door _____

4. width of a hall _____

5. length of a finger _____

6. length of an umbrella _____

7. height of a lamp _____

8. length of a shoe _____

9. width of a box _____

Using Metric Units of Mass (Weight)

Instruction

Commonly, the term **weight** is used when we mean **mass**. The **gram** and **kilogram** are units of mass in the metric system, just as the ounce and pound are in the English system. Most packaged foods have weight indicated in grams as well as in ounces or pounds.

The basic unit of mass in the metric system is the gram. Units of mass larger or smaller are formed with the same prefixes you learned when working with length: *kilo-, hecto-, deka-, deci-, centi-,* and *milli-*.

Unit	Weight
kilogram (kg)	1,000 grams
hectogram (hg)	100 grams
dekagram (dag)	10 grams
gram (g)	1 gram
decigram (dg)	0.1 gram
centigram (cg)	0.01 gram
milligram (mg)	0.001 gram

Conversions from larger to smaller units and from smaller to larger units are made in the same way for units of mass as for units of length. See lesson two.

Example

Which is larger, 63 centigrams or 15 decigrams?

To solve, either change 63 centigrams to decigrams, or change 15 decigrams to centigrams.

63 centigrams = 6.3. decigrams

15 decigrams is larger than 63 centigrams because 15 is greater than 6.3.

Cinnamon Flavor

The Delicious High Protein Cereal

Net Wt 15 oz 425g

Exercise A   Decide which unit in each group is largest and circle it.

1. 1 kilogram, 1 gram, 1 milligram
2. 1 centigram, 1 dekagram, 1 decigram
3. 1 gram, 1 hectogram, 1 kilogram
4. 1 milligram, 1 centigram, 1 decigram
5. 1 kilogram, 1 hectogram, 1 dekagram
6. 1 gram, 9 decigrams, 10 milligrams
7. 90 centigrams, 4 grams, 5 hectograms
8. 25 decigrams, 50 decigrams, 1 gram
9. 4 grams, 1 kilogram, 8 hectograms
10. 20 grams, 3 dekagrams, 100 centigrams

Exercise B   Rank in order of heaviest weight to lightest weight. Write the letters in the blanks.

1. ....   a. 1 dekagram

2. ....   b. 1 gram

3. ....   c. 1 hectogram

4. ....   d. 1 centigram

5. ....   e. 1 kilogram

6. ....   f. 1 decigram

7. ....   g. 25 decigrams

8. ....   h. 2 kilograms

9. ....   i. 5 hectograms

10. ....   j. 8 decigrams

11. ....   k. 2 centigrams

12. ....   l. 90 centigrams

13. ....   m. 50 decigrams

14. ....   n. 28 milligrams

15. ....   o. 50 hectograms

16. ....   p. 8 kilograms

14

Comparing English and Metric Units of Mass (Weight)

Instruction

It is not a good idea to try to convert metric units to English units. However, you can use conversion to get an idea of the relative size of the units. Since the comparisons will not be exact, we will use $\approx$ to mean they are approximately the same.

Example

> 1 ounce $\approx$ 28 grams
> 1 pound $\approx$ 450 grams
> 2.2 pounds $\approx$ 1 kilogram

Example

> How much would a 4-pound bag of potatoes weigh in kilograms?
>
> 2.2 pounds $\approx$ 1 kilogram
>
> $$\begin{array}{r} 1.8 \\ 2.2\overline{)4.0} \\ -22 \\ \hline 180 \\ -176 \\ \hline 4 \end{array}$$
>
> 4 pounds $\approx$ 1.8 kilograms

Example

> You have 9 ounces of grass seed. How many grams do you have?
>
> 1 ounce $\approx$ 28 grams
>
> $$\begin{array}{r} 28 \\ \times 9 \\ \hline 252 \end{array}$$
>
> 9 ounces $\approx$ 252 grams

Exercise A

Convert these measures.

1. 2 ounces $\approx$ .... grams

2. 3.5 ounces $\approx$ .... grams

3. 10 ounces $\approx$ .... grams

4. 15 ounces $\approx$ .... grams

5. 1 pound 2 ounces ≈ .... grams   **6.** 2.2 pounds ≈ ....kilogram

7. 4.5 ounces ≈ .... grams   **8.** 1.1 pounds ≈ .... kilogram

9. 6.6 pounds ≈ .... kilograms   **10.** 1.5 ounces ≈ .... grams

11. 8.8 pounds ≈ .... kilograms   **12.** 4 ounces ≈ .... grams

13. 6.2 ounces ≈ .... grams   **14.** 1.1 ounces ≈ .... grams

15. 4 pounds ≈ .... grams   **16.** 56 grams ≈ .... ounces

17. .4 ounce ≈ .... grams   **18.** 14 grams ≈ .... ounce

19. 4.4 pounds ≈ .... kilograms   **20.** .5 kilogram ≈ .... pounds

Wide World Photos

Exercise B   Convert these measures.

1. 4 ounces ≈ .... grams

2. 10 ounces ≈ ....grams

3. 9 ounces ≈ .... grams

4. 2.2 pounds ≈ .... kilogram

5. 1 pound ≈ .... grams

6. 1 pound 3 ounces ≈ .... grams

7. 2 pounds ≈ .... grams

8. 12 ounces ≈ .... grams

9. 48 kilograms ≈ .... pounds

10. 1,125 grams ≈ .... pounds

11. 17.6 kilograms ≈ .... pounds

12. 900 grams ≈ .... pounds

13. 112 grams ≈ .... ounces

14. 32 ounces ≈ .... grams

15. 5 kilograms ≈ .... pounds

16. 7 pounds ≈ .... grams

17

LESSON EIGHT: Using Metric Units of Capacity

Instruction

The **volume** of a container tells its size. The **capacity** tells how much it can hold. The basic unit of capacity in the metric system is the **liter**. A liter is a little more than a quart. Larger and smaller units of capacity are formed with the same prefixes you learned earlier.

Unit	Volume
kiloliter (kl)	1,000 liters
hectoliter (hl)	100 liters
dekaliter (dal)	10 liters
liter (ℓ)	1 liter
deciliter (dl)	0.1 liter
centiliter (cl)	0.01 liter
milliliter (ml)	0.001 liter

Conversions from one unit to another one are made in the same way as for units of length and mass. See lesson two to review the process.

Example

Which is smaller, 2 liters or 25 deciliters?

2 liters = 2.0. deciliters

2 liters is smaller because 20 is less than 25.

Exercise A Circle the smaller capacity in each pair.

1. 1 kiloliter or 1 hectoliter

2. 1 milliliter or 1 centiliter

3. 8 liters or 1 dekaliter

4. 2 dekaliters or 15 liters

5. 40 hectoliters or 1 dekaliter

6. 200 centiliters or 1 liter

7. 60 centiliters or 1 kiloliter

8. 10 deciliters or 10 dekaliters

9. 1 dekaliter or 9 liters

10. 2 kiloliters or 15 hectoliters

11. 89 liters or 100 centiliters

Exercise B   Circle the larger capacity in each pair.

1. 1 deciliter or 1 dekaliter
2. 1 liter or 1 deciliter
3. 1 deciliter or 1 liter
4. 200 centiliters or 1 liter
5. 500 liters or 1 hectoliter
6. 15 milliliters or 10 centiliters
7. 2 kiloliters or 1,000 milliliters
8. 59 liters or 159 deciliters
9. 6 milliliters or 16 deciliters
10. 9 liters or 1 dekaliter
11. 1 milliliter or 1 kiloliter
12. 1 centiliter or 1 milliliter
13. 45 milliliters or 2 deciliters
14. 4 deciliters or 75 centiliters
15. 250 dekaliters or 1 kiloliter
16. 99 hectoliters or 100 dekaliters
17. 10 dekaliters or 100 centiliters
18. 83 liters or 8 dekaliters
19. 896 hectoliters or 526 dekaliters
20. 15 liters or 51 centiliters
21. 2 dekaliters or 11 liters
22. 1 kiloliter or 900 liters
23. 100 milliliters or 1 dekaliter
24. 60 liters or 4 dekaliters
25. 950 liters or 1 kiloliter

Courtesy The Coca-Cola Company

LESSON NINE: Understanding Celsius Temperatures

Instruction

A **Celsius** thermometer has 100 divisions between the freezing point of water and the boiling point of water. Each unit is one degree celsius, symbolized as 1°C. Water freezes at 0°C and boils at 100°C. Study the thermometer diagram below to see the difference in Fahrenheit and Celsius temperatures.

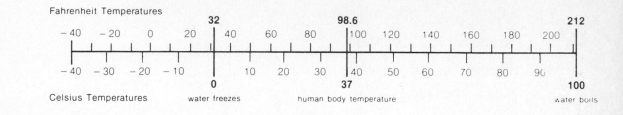

Exercise A   Solve the following problems.

1. When the temperature is 35°C, is it a good day to go swimming?

   ....

2. If your body temperature is 40°C, would you have a fever? ....

3. For which temperature would you need a coat, 25°F or 25°C? ....

4. The 2 A.M. temperatures during the week were 26°C, 24°C, 19°C, 21°C, and

   25°C. What was the average daily temperature for 2 A.M.? ....

5. The temperature dropped 3°C each hour from midnight until 6 A.M. Then it increased 2°C each hour from 6 A.M. until noon. If the midnight temperature

   was 25°C, what was the noon temperature? ....

6. A drop in temperature from 6°C to −8°C would be a drop of .... degrees.

7. You are planning a trip to a city where the average temperature is 30°C. Will

   you need heavy clothing? ....

8. What is the difference in the total number of degrees from the freezing point of water to the boiling point of water on a Fahrenheit thermometer and a Celsius

   thermometer? ....

9. At what temperature are the Fahrenheit and Celsius scales equal? ....

10. Which would you be able to drink, a glass of 100°C water or a glass of 100°F

    water? ....

Exercise B   Solve these problems.

1. At 40°C, would you be comfortable with a light coat? ....

2. When the temperature reaches 30°C, would you be able to go swimming?

   ....

3. If you measure the temperature of your hand with a Celsius thermometer, what

   should it be? ....

4. The weather report indicates that the low temperature is expected to be 3°C. Is

   that freezing? ....

5. If the thermostat in your home had a Celsius scale, would you be comfortable

   if you set the thermostat at 65°C in the winter? ....

6. Do you think you would like to go swimming in 88°C water? ....

7. An increase in temperature from 10°C to 13°C would be an increase of

   .... degrees.

8. Which is warmer, 50°F or 50°C? ....

9. The high temperature recorded one summer in a Texas city was 43°C. The low
   temperature during the following winter was −21°C. What was the difference

   between the high and low temperatures? ....

10. Which is human body temperature, 37°F or 37°C? ....

11. Will ice melt at 5°C? ....

12. Which temperature is better for snow skiing, 32°C or 0°C? ....

13. The weather reporter predicts a high temperature of 40°C? Will you need warm

    clothing? ....

14. The temperature in a meeting hall is 12°C. Which will be useful, heating or air

    conditioning? ....

Courtesy Schlegel Corporation .

21

UNIT 1—REVIEW

1. Which is the larger unit?
   a. kilometer   b. millimeter   a b

2. What is the basic unit of length in the metric system?
   a. decimeter   b. meter   a b

3. One meter is more nearly equal to which?
   a. one foot   b. one yard   a b

4. Is a kilometer more or less than a mile?
   a. more than   b. less than   a b

5. What unit of metric measure would you use to measure
   the height of a person?
   a. meter   b. millimeter   a b

6. The basic metric unit of mass is what?
   a. gram   b. kilogram   a b

7. Is a gram heavier or lighter than an ounce?
   a. lighter   b. heavier   a b

8. Which is heavier?
   a. 20 centigrams   b. 2 dekagrams   a b

9. The basic metric unit for measuring capacity is what?
   a. liter   b. milliliters   a b

10. Would large capacities usually be expressed as kiloliters or
    milliliters?
    a. kiloliters   b. milliliters   a b

11. Is a liter a little more than a quart or a gallon?
    a. quart   b. gallon   a b

12. The Celsius thermometer has how many degrees between
    the freezing and boiling points of water?
    a. 100   b. 212   a b

22

Exercise B

Circle the larger (or largest) unit in each group.

1. 1 millimeter or 1 centimeter
2. 50 centigrams or 2 decigrams
3. 1 kilometer or 5 centimeters
4. 2 centiliters, 1 deciliter, 1 liter, or 2 kiloliters

Exercise C

Fill in the blanks.

1. How many decigrams equal 1 dekagram? ....

2. The basic unit of mass is a ....

3. Write the symbols for units of mass larger than a gram. ....

   ....

4. A liter contains 1,000 ml. How many ml would be in three-fourths liter?

   ....

5. Do the prefixes for capacity differ from those used for linear or mass units?

   ....

6. The basic unit for capacity is the ....

7. How many degrees are on the Celsius thermometer between the freezing and

   boiling points of water? ....

8. What would be normal body temperature in degrees Celsius? ....

9. If the temperature is 30°C in Baltimore, would it be cold or hot? ....

10. It is raining, and the temperature is 35°C. If the temperature drops 5 degrees,

    will ice form? ....

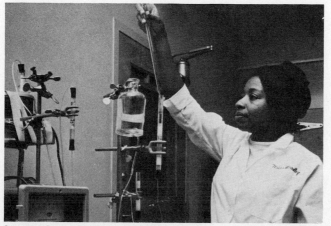

Courtesy Center for Disease Control

USING CONVERSION TABLES

During the gradual conversion from the English system of measurement to the metric system, you may find it necessary to convert measures from one system to the other. The table below lists the conversion factors for the most common measures in each system. To use the table, find the unit you know in the left-hand column and multiply it by the given number. The answer will be the approximate number of units shown in the right-hand column.

NOTE: A low cost, plastic, wallet-size metric conversion card may be purchased from the U.S. Government Printing Office, Superintendent of Documents, Washington, D.C. 20402.

WHEN YOU KNOW	MULTIPLY BY	TO FIND
Length and Distance		
inches (in.)	2.5	centimeters
inches (in.)	25	millimeters
feet (ft.)	30	centimeters
yards (yd.)	0.9	meters
miles (mi.)	1.6	kilometers
millimeters (mm)	0.04	inches
centimeters (cm)	0.4	inches
meters (m)	3.3	feet
meters (m)	1.1	yards
kilometers (km)	0.6	miles
Area		
square inches (sq. in.)	6.5	square centimeters
square feet (sq. ft.)	0.09	square meters
square yards (sq. yd.)	0.8	square meters
square miles (sq. mi.)	2.6	square kilometers
acres	0.4	hectares
square centimeters (cm^2)	0.16	square inches
square meters (m^2)	1.2	square yards
square kilometers (km^2)	0.4	square miles
hectares (ha)	2.5	acres
Capacity (Liquid)		
teaspoons (tsp.)	5	milliliters
tablespoons (T.)	15	milliliters
fluidounces (fl. oz.)	30	milliliters
cups (c.)	0.24	liters
pints (pt.)	0.47	liters
quarts (qt.)	0.95	liters
gallons (gal.)	3.8	liters
milliliters (ml)	0.03	fluidounces
liters (ℓ)	2.1	pints
liters (ℓ)	1.06	quarts
liters (ℓ)	0.26	gallons

WHEN YOU KNOW	MULTIPLY BY	TO FIND
Mass (weight)		
ounces (oz.)	28	grams
pounds (lb.)	0.45	kilograms
tons	0.9	metric tons
grams (g)	0.035	ounces
kilograms (kg)	2.2	pounds
metric tons (t)	1.1	tons
Temperature		
degrees Fahrenheit (°F)	5/9 (after subtracting 32)	degrees Celsius
degrees Celsius (°C)	9/5 (then add 32)	degrees Fahrenheit

Directions   Solve the following problems.

1. 2.5 inches ≈ .... centimeters

2. 684 millimeters ≈ .... inches

3. 9.7 meters ≈ .... feet

4. 2 liters ≈ .... pints

5. 3 tablespoons ≈ .... milliliters

6. 3 teaspoons ≈ .... milliliters

7. 16 ounces ≈ .... grams

8. 2 tons ≈ .... metric tons

9. 7.3 pounds ≈ .... kilograms

10. 17.5 metric tons ≈ .... tons

11. 2 cups ≈ .... liters

12. 4 gallons ≈ .... liters

13. 8 square feet ≈ .... square meters

14. 10.3 miles ≈ .... kilometers

15. 9 yards ≈ .... meters

16. 20 square inches ≈ .... square centimeters

Contains egg, milk and shortening

SYRUP

JUST ADD WATER

NET WT. 32 OZ. (2 LB.) 907 g   MAKES ABOUT 42—4" PANCAKES   Pancake & Waffle Mix

UNIT 2—PRETEST
SQUARES AND CUBES

Compute the perimeters of these squares.

1. $s = 89$ inches ⎯⎯⎯⎯⎯⎯⎯⎯⎯⎯

2. $s = 9.9$ meters ⎯⎯⎯⎯⎯⎯⎯⎯⎯⎯

3. $s = 8.5$ centimeters ⎯⎯⎯⎯⎯⎯⎯⎯⎯⎯

4. $s = 7\frac{1}{4}$ yards ⎯⎯⎯⎯⎯⎯⎯⎯⎯⎯

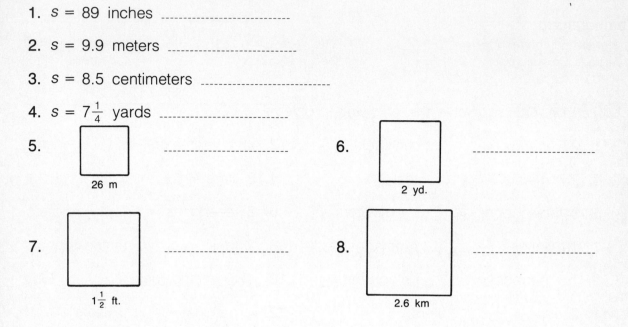

5. ⎯⎯⎯⎯⎯⎯⎯⎯⎯⎯

26 m

6. ⎯⎯⎯⎯⎯⎯⎯⎯⎯⎯

2 yd.

7. ⎯⎯⎯⎯⎯⎯⎯⎯⎯⎯

$1\frac{1}{2}$ ft.

8. ⎯⎯⎯⎯⎯⎯⎯⎯⎯⎯

2.6 km

Compute the areas of these squares.

9. $s = 2,000$ feet ⎯⎯⎯⎯⎯⎯⎯⎯⎯⎯

10. $s = 5.46$ meters ⎯⎯⎯⎯⎯⎯⎯⎯⎯⎯

11. $s = 776$ inches ⎯⎯⎯⎯⎯⎯⎯⎯⎯⎯

12. $s = .4$ centimeter ⎯⎯⎯⎯⎯⎯⎯⎯⎯⎯

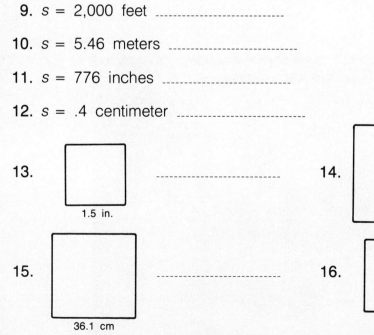

13. ⎯⎯⎯⎯⎯⎯⎯⎯⎯⎯

1.5 in.

14. ⎯⎯⎯⎯⎯⎯⎯⎯⎯⎯

16 yd.

15. ⎯⎯⎯⎯⎯⎯⎯⎯⎯⎯

36.1 cm

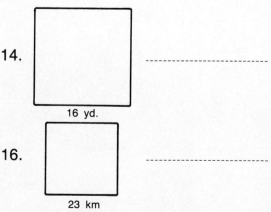

16. ⎯⎯⎯⎯⎯⎯⎯⎯⎯⎯

23 km

Compute the total surface areas of these cubes.

17. s = 1.6 meters _____

18. s = 7.3 feet _____

19. s = 19.4 millimeters _____

20. s = 8 inches _____

21. 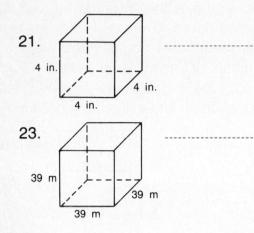 _____

22. 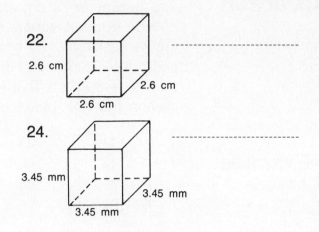 _____

23. 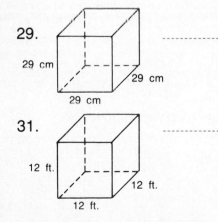 _____

24. 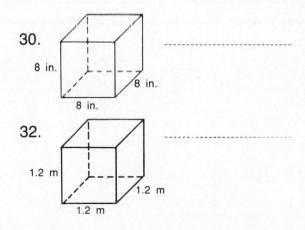 _____

Compute the volumes of these cubes.

25. s = 11 feet _____

26. s = 4.5 centimeters _____

27. s = 21.2 inches _____

28. s = 3.3 meters _____

29. _____

30. _____

31. _____

32. _____

Solve the following problems.

33. How many cubic inches of water does it take to fill a cube with 10-inch sides? _____

34. Find the area of a square patio whose perimeter is 48 feet. _____

27

SQUARES AND CUBES

Computing Perimeter of a Square

Instruction

The **perimeter** of any geometric shape is the distance around its outside edge. A **square** is a 4-sided figure where all four sides are the same length. To find the perimeter of a square, you need to find the sum of the lengths of the four sides, $P = s_1 + s_2 + s_3 + s_4$. This formula can be simplified to $P = 4s$. This means that the perimeter is equal to 4 times the length of one side.

Example

$P = 4s$
$P = 4 \times 2$
$P = 8$ inches

The perimeter of a square with 2-inch sides is 8 inches.

It is also possible to compute the length of the sides of a square when only the perimeter is known. This is done by rearranging the formula to look like this: $s = \frac{P}{4}$. This means one side is equal to the perimeter divided by 4.

Example

The perimeter of a square room is 36 feet. What is the length of each side of the room?

$s = \frac{P}{4}$

$s = \frac{36}{4}$

$s = 9$ feet   Each side of the room is 9 feet long.

Exercise A   Solve the following problems.

1. Find the perimeter of this square.   -----------------

36 cm

2. To build two forms for two identical square buildings, how many linear feet of $2'' \times 12''$ lumber will be required if $s = 40$ feet for each?   -----------------

28

3. What size painting may be purchased to fit a square antique picture frame with an inside perimeter of 56 inches?

----

4. Chen has a garden spot that measures 100 feet on each side. How much will it cost to fence it if the fence costs $2.86 per foot?

----

5. The distance between each base on a baseball diamond is 90 feet. How far would you have jogged if you made 9 trips around the diamond?

----

6. You are building a house with 12 windows. Each window is 4 feet square. How many feet of lumber would you need to place a border around the windows?

----

7. Lawrence wants to fence a piece of land that is 1 mile long on each side. How many rolls of barbed wire will he need if he wants the fence to have 3 strands of wire on each side? (A roll of barbed wire is $\frac{1}{4}$ mile in length.)

----

8. What would be the dimensions of a square patio having a perimeter of 40 kilometers?

----

9. Three squares have a combined perimeter of 144 feet. The smaller one has 4-foot sides. The second square is three times as large as the first. The third square is five times as large as the first. Find the dimensions of each square.

----

10. Find the perimeter of a square with sides that are 5 inches long.

----

11. What is the perimeter of a square whose sides are $\frac{2}{3}$ inch long?

----

12. If each side of a square is made twice its original size, what change takes place in its perimeter?

Exercise B   Solve these problems.

1. A square has a perimeter of 328 feet. What is the length of each side?   ....

2. A city lot is 90 meters square. What is the perimeter of the lot?   ....

3. How many fence posts would be required to fence a parcel of land that is $\frac{1}{2}$ mile square if the posts are placed 8 feet apart. (1 mile = 5,280 feet.)   ....

4. Anthony has 128 feet of string to mark the perimeter of his garden. What must the length of each side be if he makes the garden square?   ....

5. What is the length of one side of a square when the perimeter is 36 feet?   ....

6. Find the length of one side of a square having a perimeter of 144 meters.   ....

7. A local shop charges $3.12 per foot to place a frame around a square mirror with 24-inch sides. What will the frame cost?   ....

8. How many meters of neon tubing will be needed to put a border around a square sign with sides that are 1.2 meters long?   ....

9. A square has a perimeter of 132 meters. What is the length of each side?   ....

10. Mrs. Swenson needs to have a vacant lot fenced. How much wire does she need to buy if the lot measures 84 feet on each side?   ....

LESSON TWO: Computing the Area of a Square

Instruction

The surface enclosed by the borders of a plane figure is called the **area** of the figure. The area of any figure is the number of **square units** it contains.

Example

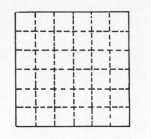

This space can be divided into 36 square units. Its area is 36 square units.

To find the **area** of a **square** you can use the formula $A = s^2$. This formula is read "area equals side squared." That means to multiply the length of the side times itself.

Example

Find the area of a square with a side that is 7 inches long.

$A = s^2$
$A = 7 \times 7$
$A = 49$ square inches

Exercise A   Solve the following problems.

1. How many square feet are in a square that measures 3 feet by 3 feet?   ....

2. Find the area of a square having a side of 12 centimeters.   ....

3. How much more area does a square with 4-foot sides have than one with 2-foot sides?   ....

4. To paint a square that is 8 meters on each side will require how many pints of paint if one pint will cover 4 square meters?   ....

5. Find the area of a square with a side that is 3 yards and 2 feet long. (Note, the length of the side is given in mixed units. To solve, you need to convert the length to either yards or feet.)   ....

6. How many yards of carpet are needed to cover a room measuring 20 feet by 20 feet? (1 square yard contains 9 square feet.)

....

7. If a 2-inch square picture is enlarged to 3 times its original size, how many square inches are in the enlarged picture?

....

8. How many more square inches are in the enlarged picture than in the original?

....

Exercise B   Solve the following problems.

1. The Stubblefield's yard is square, and one side is 100 feet long. They wish to kill the weeds on the lawn. If one pound of weed killer covers 200 square feet, how many pounds will be needed to do the entire yard?

....

2. Floor covering is on sale for $3.95 per square yard. What will be the cost of covering a room that measures 12 feet by 12 feet?

....

3. Find the area of a workshop that measures 22 feet by 22 feet.

....

4. Jennifer wishes to cover one wall in her bathroom with wallpaper. The wall is square and is 8 feet high. How many square feet of wallpaper will she need for the job?

....

5. How many 1-centimeter squares can be placed in a square that measures 32 centimeters on each side?

....

6. The warehouse is built on a lot that is 700 feet square. What is the area of the lot?

....

LESSON THREE: Find the Total Surface Area of a Cube

Instruction To find the total surface area of a **cube**, first find the area of one side. Then multiply by 6. Thus, the formula for the total surface area of a cube is $T = 6s^2$.

Example

The sides of a cube are 2 feet long. Find the total surface area of the cube.

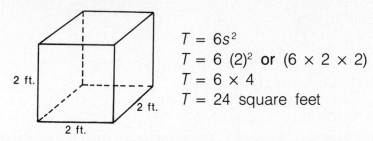

2 ft.

2 ft.

2 ft.

$T = 6s^2$
$T = 6 \ (2)^2$ **or** $(6 \times 2 \times 2)$
$T = 6 \times 4$
$T = 24$ square feet

Think, a cube has 6 sides, and each side is 2 feet long. Thus, $2 \times 2 =$ the area of one side. Multiplying 6 times the area of one side gives the total surface area of the cube.

Exercise A Find the total surface area of these cubes.

1. $s = 1$ foot ....

2. $s = 5$ yards ....

3. $s - 6$ feet 6 inches

4. $s = 7.3$ inches ....

5. $s = 9$ inches ....

6. $s = 1.5$ meters ....

7. $s = 1.6$ inches ....

8. $s = .25$ centimeters ....

9. $s = 12$ feet ....

10. $s = 2.75$ feet ....

11. $s = 10$ meters ....

12. $s = 3$ inches ....

13. $s = 14.6$ millimeters ....

Exercise B   Find the total surface areas of these cubes.

1. s = 15 yards ....

2. s = 8 centimeters ....

3. s = 1 inch ....

4. s = 2 feet 4 inches ....

5. s = 3 yards 1 foot ....

6. s = $1\frac{1}{3}$ yards ....

7. s = $2\frac{1}{2}$ feet ....

8. s = 24 feet ....

9. s = 15 inches ....

10. s = 12.8 centimeters ....

11. s = 4 feet ....

12. s = 2 meters ....

13. s = 7 inches ....

14. s = 6.1 centimeters ....

15. s = 4 inches ....

16. s = 4 meters ....

17. s = $1\frac{1}{2}$ feet ....

18. s = 1 meter ....

19. s = 3.2 inches ....

20. s = $8\frac{1}{2}$ millimeters ....

21. s = 12 centimeters ....

22. s = 15 feet ....

23. s = 2 yards ....

24. s = 5 inches ....

25. s = 7 yards ....

26. s = 6 feet ....

27. s = 2.4 meters ....

LESSON FOUR: Finding the Volume of a Cube

Instruction

The space enclosed within a three-dimensional figure is called **volume**. Volume is measured in **cubic units**. The formula for finding the volume of a cube is $V = s^3$. It is read "volume equals side cubed." To cube a number, multiply it times itself two times.

Example

Find the volume of a cube that has 2-inch sides.

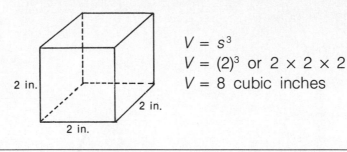

2 in.
2 in.
2 in.

$V = s^3$
$V = (2)^3$ or $2 \times 2 \times 2$
$V = 8$ cubic inches

Exercise A   Solve the following problems.

1. A bin that measures 16 feet on each side is filled with grain. How many bushels does it contain? (1.25 cu. ft. = 1 bu.)   ....

2. What is the volume of a cube when $s = 10$ yards?   ....

3. Which is larger, a 4-inch cube or 4 cubic inches?   ....

4. How many gallons of water will a cube that is 4 feet on each side hold? ($7\frac{1}{2}$ gallons = 1 cubic foot.)   ....

5. What is the volume of a cube with sides that are 6 feet 6 inches long?   ....

6. Express in cubic inches the volume of a cube with 8-inch sides.   ....

7. What is the volume of a cube having 16-centimeter sides?   ....

8. Delfina is building a sandbox for her children. Each side is 3 feet long, and the box is 3 feet deep. How many cubic feet of sand will it hold?   ....

9. Compute the volume of a cube with sides 6.3 centimeters long.   ....

Exercise B   Find the volumes of these cubes.

1. s = 9.5 feet ....

2. s = 15 meters ....

3. s = 5 feet 6 inches ....

4. s = 9 inches ....

5. s = 23 feet ....

6. s = 2 feet 4 inches ....

7. s = 53 inches ....

8. $s = 3\frac{1}{3}$ yards ....

9. s = 7 centimeters ....

10. s = 27.6 feet ....

11. s = 4 inches ....

12. s = 10 centimeters ....

13. s = 1 foot ....

14. $s = 3\frac{1}{2}$ feet ....

15. s = 4 meters ....

16. s = 20 feet ....

17. s = 31 centimeters ....

18. $s = 4\frac{1}{2}$ centimeters ....

19. s = 1 foot 8 inches ....

20. s = 7 millimeters ....

21. s = 4 feet ....

22. s = 2 inches ....

23. s = 5 centimeters ....

24. s = 6 yards ....

25. s = 2 feet ....

26. $s = 1\frac{1}{2}$ inches ....

27. s = 4.2 centimeters ....

UNIT 2—REVIEW

Exercise A   Compute the perimeters of these squares.

1. s = 106 feet 4 inches ------------------------------

2. s = 96 inches ------------------

3. $s = 8\frac{1}{3}$ yards ------------------

4. s = 2.7 meters ------------------

5. s = 3.4 centimeters ------------------

Exercise B   Compute the areas of these squares.

1. s = 16 inches ------------------

2. s = 1,320 feet ------------------

3. s = 27.6 meters ------------------

4. s = 1.43 dekameters ------------------

5. s = 2.6 meters ------------------

Exercise C   Compute the total surface areas of these cubes.

1. s = 6 feet ------------------

2. s – 4 inches

3. s = 1.83 centimeters ------------------

4. s = 29.9 meters ------------------

5. s = 13 inches ------------------

Exercise D   Compute the volumes of these cubes.

1. s = 2.6 inches ------------------

2. s = 13 feet ------------------

3. s = 47.6 centimeters ------------------

4. s = 5.4 meters ------------------

5. s = 6.2 centimeters ------------------

Exercise E   Solve the following problems.

1. How many feet of lumber will be needed to make a form for a patio that is to be 21 feet square?

    ----

2. To place a fence around a garden that is 56 feet square would require how many feet of wire?

    ----

3. If a square picture has a perimeter of 48 inches, what is the length of each side?

    ----

4. Kirby cut a square opening in the ceiling of his carport. If one side measures 48 inches, how much 1″ × 4″ lumber will be needed to go around the opening?

    ----

5. What is the surface area of a square whose sides measure 52 inches?

    ----

6. Jeremy is going to make a tablecloth for a table. The table is 72 inches square. If he wishes to have a 6-inch overhang, what will be the total area of the tablecloth?

    ----

7. What is the difference in the area of a square having a side of 3 inches and one having 9-inch sides?

    ----

8. Find the cost, at $5.50 per square foot, of laying a concrete patio that measures 14 feet square.

    ----

9. What is the area of a garden that measures 35 feet on each side?

    ----

10. What is the volume of a cube having 5-centimeter sides?

Courtesy Armstrong Cork Company

38

READING LAND SECTION MAPS

Quite often disputes arise between people concerning the boundaries of real estate. A common reason for this is that people do not know how to figure the boundaries of their property from the description given in their deed. A diagram of a section of land is shown below.

Note that a section of land is one square mile in perimeter and has an area of 640 acres. It is further divided into fourths, with each one-fourth containing 160 acres. Like any map, the direction of north is to the top, south to the bottom, east to the right, and west to the left.

The map shows a piece of land referred to as Section 6. A legal description of a portion of the section might read as follows: NW $\frac{1}{4}$ of Sec. 6. This would simply be $\frac{1}{4}$ of the section. It would be the fourth in the upper left-hand corner of the section, and it would contain 160 acres.

Another 40-acre parcel of land is located in the NE $\frac{1}{4}$ of the SE $\frac{1}{4}$ of Sec. 6. To find this section, locate the SE $\frac{1}{4}$ of section 6. Then subdivide that $\frac{1}{4}$ of the section into fourths and then locate the NE $\frac{1}{4}$ of the SE $\frac{1}{4}$. This would be the portion located in the upper right-hand corner of the SE $\frac{1}{4}$.

Directions

Locate a 10-acre parcel of land in the NE $\frac{1}{4}$ of the SE $\frac{1}{4}$ of the SW $\frac{1}{4}$ of Sec. 6. Shade that portion of land. What is its perimeter in feet? (Hint: Start by locating the last description and work backwards.)

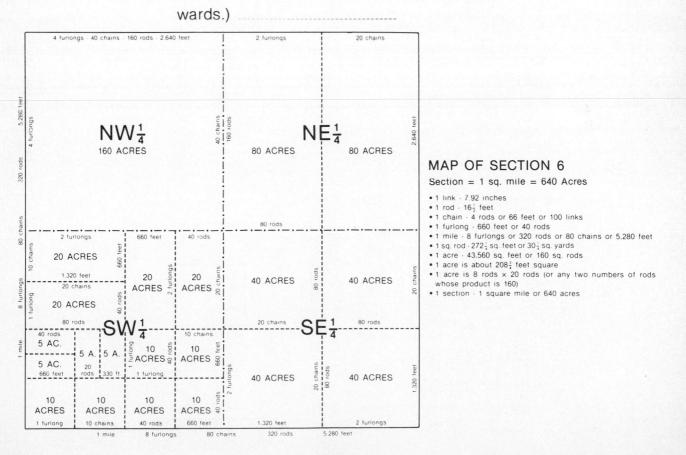

MAP OF SECTION 6

Section = 1 sq. mile = 640 Acres

- 1 link · 7.92 inches
- 1 rod · 16$\frac{1}{2}$ feet
- 1 chain · 4 rods or 66 feet or 100 links
- 1 furlong · 660 feet or 40 rods
- 1 mile · 8 furlongs or 320 rods or 80 chains or 5,280 feet
- 1 sq. rod · 272$\frac{1}{4}$ sq. feet or 30$\frac{1}{4}$ sq. yards
- 1 acre · 43,560 sq. feet or 160 sq. rods
- 1 acre is about 208$\frac{1}{2}$ feet square
- 1 acre is 8 rods × 20 rods (or any two numbers of rods whose product is 160)
- 1 section · 1 square mile or 640 acres

UNIT 3—PRETEST
RECTANGLES AND RECTANGULAR SOLIDS

Find the perimeters of these rectangles.

1. l = 16 feet; w = 10 feet ----------------------------

2. l = 21 inches; w = 14.5 inches ----------------------------

3. l = 40 yards; w = 27 yards ----------------------------

4. l = 24.3 meters; w = 36.2 meters ----------------------------

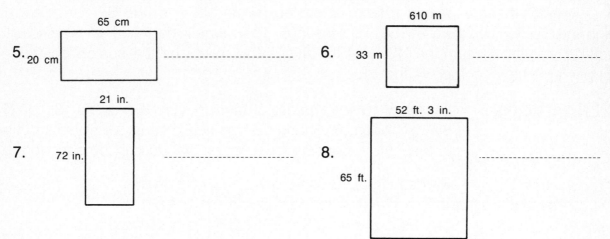

5. 65 cm / 20 cm ----------------------------

6. 610 m / 33 m ----------------------------

7. 21 in. / 72 in. ----------------------------

8. 52 ft. 3 in. / 65 ft. ----------------------------

Find the areas of these rectangles.

9. l = 11 feet; w = $8\frac{1}{2}$ feet ----------------------------

10. l = 4 meters; w = 2.4 meters ----------------------------

11. l = 2.6 centimeters; w = 2.4 centimeters ----------------------------

12. l = 14 inches; w = 11 inches ----------------------------

13. 5.2 ft. / 1 ft. ----------------------------

14. 7.1 in. / 4.1 in. ----------------------------

15. 11 m / 23 m ----------------------------

16. $3\frac{1}{2}$ cm / 2.1 cm ----------------------------

Find the volumes of these rectangular solids.

17. l = 7 inches; w = 5 inches; h = $1\frac{1}{2}$ inches. _____

18. l = 3 feet; w = $1\frac{1}{4}$ feet; h = $1\frac{1}{4}$ feet _____

19. l = 4 meters; w = 2 meters; h = 1.5 meters _____

20. l = 42 centimeters; w = 21 centimeters; h = 12 centimeters _____

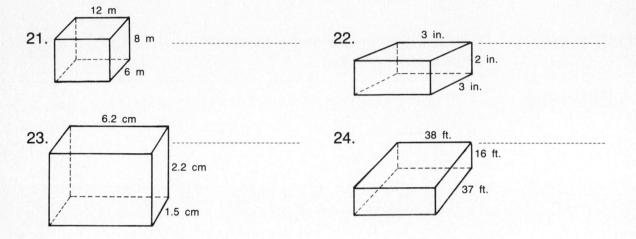

21. 12 m / 8 m / 6 m _____

22. 3 in. / 2 in. / 3 in. _____

23. 6.2 cm / 2.2 cm / 1.5 cm _____

24. 38 ft. / 16 ft. / 37 ft. _____

Solve the following problems. Blacken the letter to the right that corresponds to the letter before each correct answer.

25. Janette built a pen for her pets that measures 5 meters by 5.5 meters. How many meters of wire did she use?
a. 21 meters   b. 19 meters   c. 22 meters   d. 27 meters   ⓐ ⓑ ⓒ ⓓ

26. Which rectangle is larger?
a. 10 feet by 12 feet   b. 11 feet by 11 feet   ⓐ ⓑ

27. How many more square inches are in a rug measuring 48 inches by 64 inches than 36 inches by 48 inches?
a. 1,244 sq. in.   b. 1,144 sq. in.   c. 1,344 sq. in. d. 1,444 sq. in.   ⓐ ⓑ ⓒ ⓓ

28. The Hardaways' backyard measures 50 feet by 75 feet. If it takes 6 pounds of fertilizer to fertilize 1,000 square feet of lawn, how many 25-pound sacks should the Hardaways buy?
a. two sacks   b. one sack   c. three sacks   d. ten sacks   ⓐ ⓑ ⓒ ⓓ

29. Figure the perimeter of a canvas measuring 16 inches by 2 feet.
a. 4 feet 32 inches   b. 36 inches   c. 7 feet   d. 80 inches

RECTANGLES AND RECTANGULAR SOLIDS

LESSON ONE: Computing the Perimeter of a Rectangle

Instruction

To find the **perimeter** of any **rectangle**, you can use the following formula: $P = 2l + 2w$.

Example

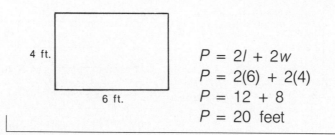

Find the perimeter of the following rectangle.

4 ft.

6 ft.

$P = 2l + 2w$
$P = 2(6) + 2(4)$
$P = 12 + 8$
$P = 20$ feet

Exercise A

Solve the following problems.

1. How much wire will Francis need to fence a rectangular garden that is 300 feet by 160 feet?

2. Find the perimeter of a picture that measures 8 inches by 10 inches.

3. What will it cost to replace molding placed around a floor that is 12 feet long and 10 feet 6 inches wide if the molding sells for 23¢ a foot?

4. Mr. Scardino bought braid to put around a pillow that measures 21 inches by 24 inches. How much did he pay for the braid at 40 cents a yard?

5. A hardware store advertised chain link fence at $2.75 per foot. What would it cost to fence a lot 100 feet by 150 feet if an opening 8 feet wide is left for a driveway?

6. Delfina's front yard is 70 feet long and 35 feet wide. What is the perimeter of her front yard?

Exercise B   Find the perimeters of these rectangles.

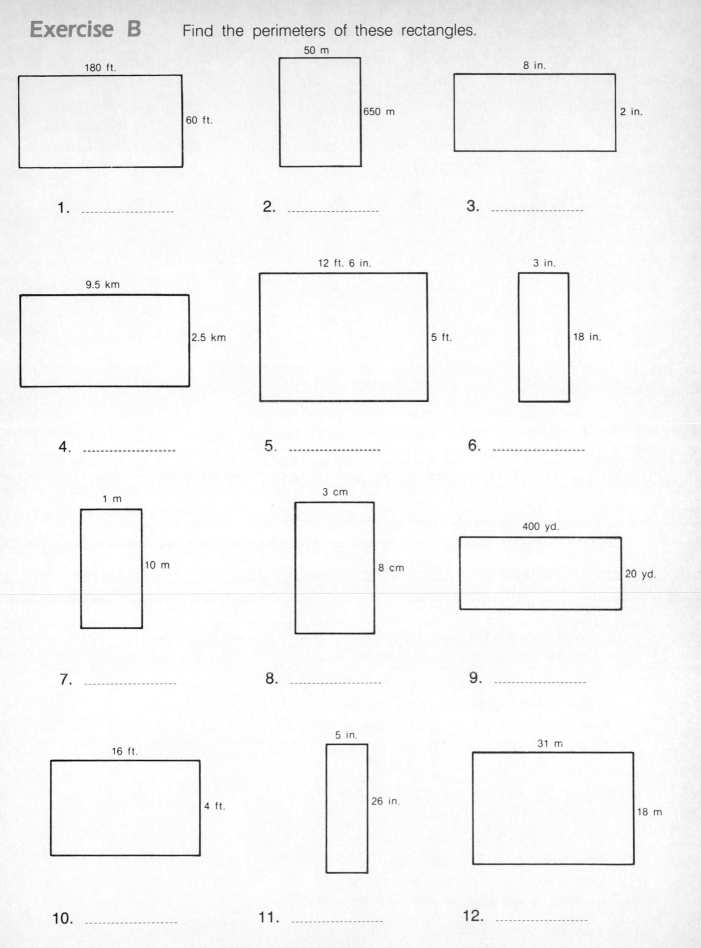

180 ft.

60 ft.

1. _____

50 m

650 m

2. _____

8 in.

2 in.

3. _____

9.5 km

2.5 km

4. _____

12 ft. 6 in.

5 ft.

5. _____

3 in.

18 in.

6. _____

1 m

10 m

7. _____

3 cm

8 cm

8. _____

400 yd.

20 yd.

9. _____

16 ft.

4 ft.

10. _____

5 in.

26 in.

11. _____

31 m

18 m

12. _____

43

LESSON TWO: Finding the Area of a Rectangle

Instruction

The **area** of a **rectangle** is the number of square units it contains. To find the area, use this formula: $A = lw$. Note, both the length and width of the rectangle must be in the same unit before multiplying.

Example

Find the area of a room that measures 12 feet by 10 feet 6 inches.

10 feet 6 inches = $10\frac{1}{2}$ feet = 10.5 feet.

$A = lw$
$A = 12 \times 10.5$
$A = 126$ square feet

Exercise A Solve the following problems.

1. How many yards of carpet will be required to carpet a room 12 feet by 14 feet? (1 sq. yd. = 9 sq. ft.)   ....

2. Marcus had his living room carpet cleaned at a cost of 9 cents per square foot. The carpet measures 15 feet by 11 feet. How much did he pay?   ....

3. Which is larger, a room measuring 9 feet by 11 feet or one measuring 8 feet by 12 feet. How much larger is it?   ....

4. To sod a yard 100 feet by 60 feet would require how many square feet of sod?   ....

5. Jerome has a rectangular plot of ground that measures 6 meters by 12 meters. He wishes to divide it into equal rectangles that measure 3 meters by 4 meters. How many rectangles will he have?   ....

6. A legal pad measures $8\frac{1}{2}$ inches by 14 inches. What is the area of one sheet?   ....

7. A fertilizer company recommends 6 pounds of fertilizer per 1,000 square feet of lawn. How much fertilizer will be required to fertilize an area 75 feet by 125 feet?   ....

8. A painter needs to paint a wall that is 12 feet long and 8 feet high. How many square feet will she paint?

....

Exercise B   Solve the following problems.

1. How many pints of paint are needed to paint a wall 8 feet by 20 feet if a pint will cover 40 square feet?

....

2. What is the area of a sidewalk measuring 3 feet by 40 feet?

....

3. How many more square inches are in a picture measuring 8″ × 12″ than one measuring 4″ × 8″?

....

4. The Andrianis' new house is 30 feet wide and 60 feet long. How many square feet are in the house?

....

5. An artist agreed to paint a landscape picture for Miguel at 20¢ per square inch. How much will it cost for a picture 15 inches by 16 inches?

....

6. What is the area of a garden measuring $9\frac{1}{2}' \times 11'$?

....

7. How many square yards of carpet are needed to cover a room measuring 15′ × 20′?

....

8. A copy machine reduces an $8\frac{1}{2}'' \times 11''$ document to 76% of its original size. What is the area of the copy?

....

45

Finding the Volume of Rectangular Solids

Instruction

To find the volume of a rectangular solid (box), multiply the length × width × height. The formula is $V = lwh$. Remember, volume is expressed in cubic units.

Example

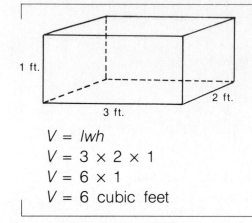

1 ft.

3 ft.

2 ft.

This box measures 3 feet by 2 feet by 1 foot. Find its volume.

$V = lwh$
$V = 3 \times 2 \times 1$
$V = 6 \times 1$
$V = 6$ cubic feet

Exercise A   Solve the following problems.

1. How many loads of dirt are required to fill a hole 40 feet long, 27 feet wide, and 5 feet deep if each load contains 5 cubic yards? (1 cu. yd. = 27 cu. ft.)   ....

2. A rectangular tank measuring 40 feet by 6 feet is filled with water 1 foot from the top. How many gallons of water are in the tank if the tank is 8 feet deep? ($7\frac{1}{2}$ gallons = 1 cubic foot)   ....

3. Find the volume of a box that is 5 feet long, 2 feet wide, and 3 feet high.   ....

4. A filing cabinet drawer is $1\frac{1}{2}$ feet wide, $1\frac{1}{2}$ feet deep, and 3 feet long. What is its volume?   ....

5. How many boxes that are 1 foot on every side will fit into a room 10 feet high, 20 feet long, and 12 feet wide?   ....

6. What is the volume of a box 1 yard long, 2 feet wide, and 10 inches high?   ....

7. Compute the volume of a trash bin that measures 1.5 meters by 2 meters by 4 meters.   ....

8. Maxwell has two rectangular boxes. One measures 2 feet by 6 inches by 10 feet. The sides of the second box are $\frac{1}{2}$ the size of the first one. What is the combined volume of the two boxes?

----

9. How much space is in an ice chest with inside dimensions of 3 feet by 18 inches by 2 feet?

----

10. Find the volume of a box that is 10 inches long, 8 inches wide, and 6 inches deep.

Exercise B   Solve the following problems.

1. Compute the volume of a room that is 24 feet long, 22 feet wide, and 12 feet high.

----

2. A cubic foot of water weighs 62.5 pounds. How much does the water in a tank that is 10 feet by 2 feet weigh if it is filled to a depth of 6 inches?

----

3. A warehouse measures 40 meters by 30 meters by 14 meters. How many cubic meters of space are in the warehouse?

----

4. Find the volume of a rectangular box measuring 3 centimeters by 5 centimeters by 8 centimeters.

----

5. How many cubic feet of space will Karen have in her workshop if it measures 10 feet by 12 feet by 8 feet, and she uses $\frac{1}{3}$ of the space to store tools?

----

6. A building measures 56 feet by 23 feet by 12 feet. Carlos wishes to rent $\frac{1}{4}$ of the space. He must pay .83¢ per cubic foot per year. What will his rent be?

UNIT 3—REVIEW

Exercise A   Find the perimeters of these rectangles.

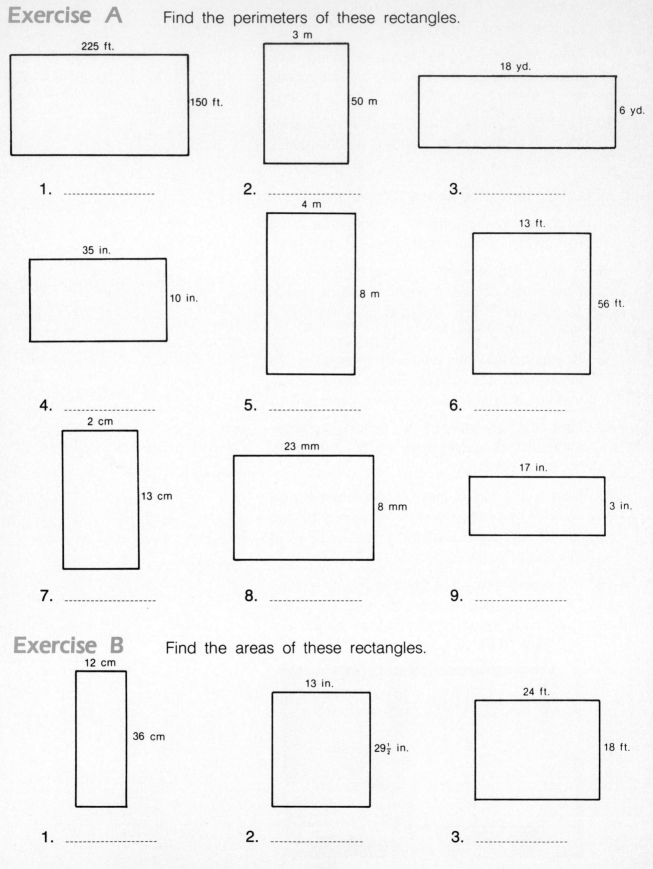

225 ft.

150 ft.

1. _____

3 m

50 m

2. _____

18 yd.

6 yd.

3. _____

35 in.

10 in.

4. _____

4 m

8 m

5. _____

13 ft.

56 ft.

6. _____

2 cm

13 cm

7. _____

23 mm

8 mm

8. _____

17 in.

3 in.

9. _____

Exercise B   Find the areas of these rectangles.

12 cm

36 cm

1. _____

13 in.

$29\frac{1}{2}$ in.

2. _____

24 ft.

18 ft.

3. _____

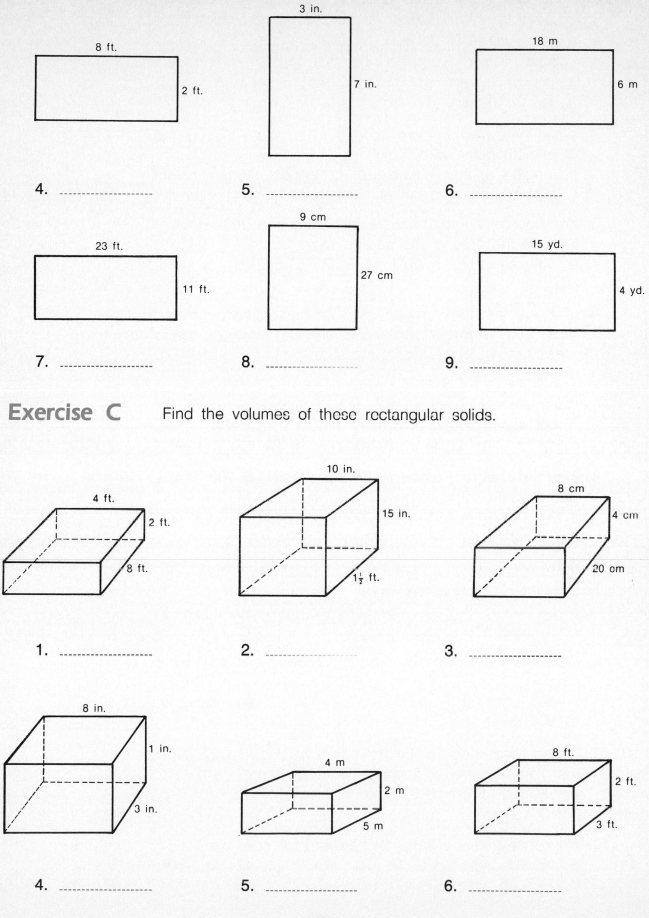

8 ft.

2 ft.

4. _____

3 in.

7 in.

5. _____

18 m

6 m

6. _____

23 ft.

11 ft.

7. _____

9 cm

27 cm

8. _____

15 yd.

4 yd.

9. _____

Exercise C   Find the volumes of these rectangular solids.

4 ft.

2 ft.

8 ft.

1. _____

10 in.

15 in.

$1\frac{1}{2}$ ft.

2. _____

8 cm

4 cm

20 cm

3. _____

8 in.

1 in.

3 in.

4. _____

4 m

2 m

5 m

5. _____

8 ft.

2 ft.

3 ft.

6. _____

49

Exercise D

Solve the following problems. Blacken the letter to the right that corresponds to the letter before each correct answer.

1. What is the formula for finding the perimeter of a rectangle?
   a. $P = 2l + 2w$ b. $P = 3l + 3w$ c. $P = 3l + 2w$
   d. $P = 2l + 3w$

   a b c d

2. What measurements are needed to compute the volume of a rectangular solid (box)?
   a. length, width, diameter   b. length, width, height
   c. length, height, diameter   d. width, height, diameter

   a b c d

3. The floor of a garage is 22 feet wide and 26 feet long. What is the area of the garage floor?
   a. 467 sq. ft.   b. 573 sq. ft.   c. 562 sq. ft.   d. 572 sq. ft.

   a b c d

4. Is it true that the formulas used for computing perimeter, area, and volume of rectangles are the same for metric measures as for traditional measures?
   a. yes   b. no

   a b

5. Compute the perimeter of a rectangle having a width of 4 feet and a length of 6 feet.
   a. 12 ft.   b. 20 ft.   c. 30 ft.   d. 24 ft.

   a b c d

6. Find the area of a rectangle having a length of 100 meters and a width of 30 meters.
   a. 3,000 sq. meters   b. 300 sq. meters   c. 30,000 sq. meters   d. 30 sq. meters

   a b c d

7. What is the volume of a box that measures 5 feet long, 2 feet wide, and 3 feet high?
   a. 36 cubic ft.   b. 46 cubic ft.   c. 30 cubic ft.
   d. 3 cubic ft.

   a b c d

8. Compute the volume of a room 12 feet by 10 feet by 8 feet.
   a. 690 cubic ft.   b. 860 cubic ft.   c. 960 cubic ft.
   d. 900 cubic ft.

   a b c d

9. A den measuring 18′ by 21′ has a hardwood floor. How many square feet of flooring are in the room?
   a. 360 sq. ft.   b. 378 sq. ft.   c. 387 sq. ft.   d. 3,780 sq. ft.

   a b c d

10. How many square yards of hardwood floor is that?
    a. 18 sq. yds.   b. 16 sq. yds.   c. 42 sq. yds.
    d. 18.5 sq. yds.

    a b c d

BUYING BUILDING MATERIALS

Building materials are bought in certain lengths. Usually, the length is in multiples of 2. For example, a sheet of plywood is 4 feet wide and 8 feet long. The same plywood can also be bought in 10 feet and 12 feet lengths, but it cannot be found in 7 feet, 9 feet, or 11 feet lengths. Therefore, it must be bought in dimensions larger than needed so it can be cut to the desired size. These facts concerning building materials can allow you to have a larger house, room, or whatever if you plan them to be built according to standard sizes of materials.

An architect will design a building for you and take advantage of these facts. To get the most for your money, consult an architect to help you design the blueprints for your building.

Directions   You wish to panel a room that measures 11 feet by 8 feet and is 8 feet high. How many sheets of paneling will be needed to panel the walls if you buy 4 feet by 8 feet paneling? Do not make any allowances for windows or doors. _____

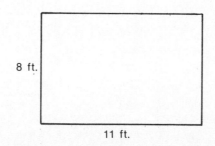

8 ft.

11 ft.

UNIT 4—PRETEST
TRIANGLES

Label these angles as acute, right, obtuse, straight, or reflex.

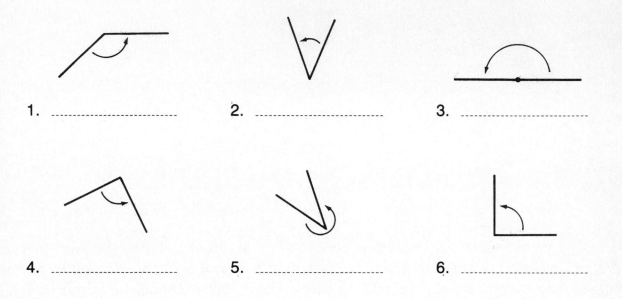

1. _____   2. _____   3. _____

4. _____   5. _____   6. _____

Match the types of angles to their descriptions. Write the correct letter in each blank.

7. _____ right   a. an angle of 180 degrees

8. _____ acute   b. an angle of 90 degrees

9. _____ obtuse   c. an angle of less than 90 degrees

10. _____ straight   d. an angle of more than 90 degrees but less than
                               180 degrees

11. _____ reflex

                            e. an angle of more than 180 degrees but less than
                               360 degrees

Find the areas of these triangles.

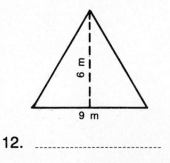

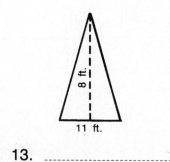

12. _____   13. _____

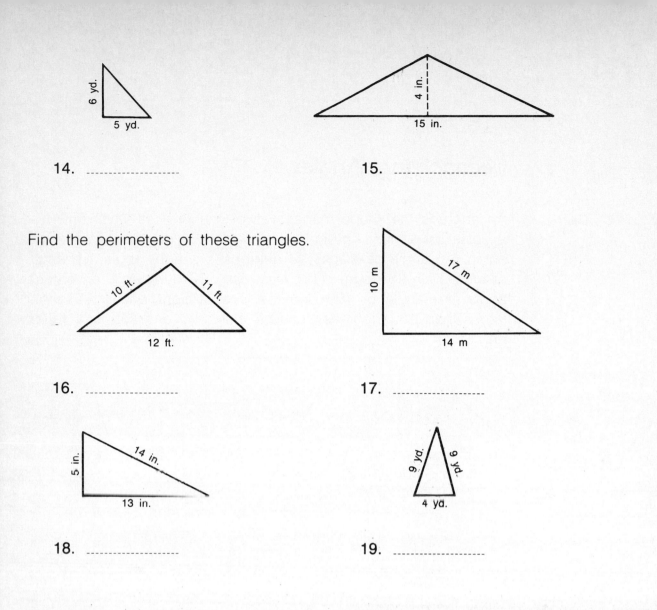

14. ....

15. ....

Find the perimeters of these triangles.

16. ....

17. ....

18. ....

19. ....

Read and solve these problems. Blacken the letter to the right that corresponds to the letter before each correct answer.

20. Find the area of a triangular plot of land. The base of the triangle is 3.6 miles and the altitude is 1.2 miles.
a. 21.6 miles   b. 2.16 miles   c. 216 miles   d. 2.26 miles

[a] [b] [c] [d]

21. A triangular pane of glass has a base of 20 inches and an altitude of 18 inches. What is the area of the pane of glass?
a. 180 sq. in.   b. 18 sq. in.   c. 18.5 sq. in.   d. 80 sq. in.

[a] [b] [c] [d]

22. A school club makes pillows to sell to finance its projects. The pillows measure 18 inches on each side and are edged with fringe. This year the club plans to produce 2,000 of the pillows. How many yards of fringe do they need to order for this year's pillows?
a. 108,000   b. 2,000   c. 54   d. 3,000

[a] [b] [c] [d]

53

TRIANGLES

Recognizing Angles

Instruction

An **angle** is the shape made by two straight lines which meet at a point, the vertex. An angle of less than 90 degrees is an **acute** angle. An angle of exactly 90 degrees is a **right** angle. An angle of more than 90 degrees but less than 180 degrees is an **obtuse** angle. An angle of 180 degrees is a **straight** angle. An angle larger than 180 degrees but less than 360 degrees is a **reflex** angle.

Example

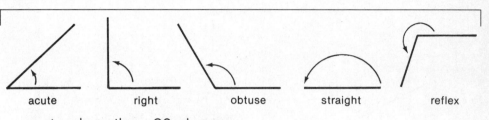

acute   right   obtuse   straight   reflex

acute—less than 90 degrees
right—an angle of exactly 90 degrees
obtuse—more than 90 degrees but less than 180 degrees
straight—an angle of exactly 180 degrees
reflex—larger than 180 degrees but less than 360 degrees

Exercise A   Label these angles.

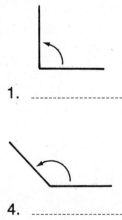

1. _____

2. _____

3. _____

4. _____

5. _____

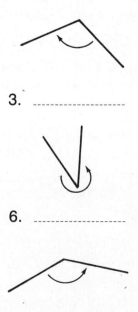

6. _____

7. _____

8. _____

9. _____

Exercise B   Several angles are described below. Write the name of each angle on the line and then draw an example of the angle.

1. .... an angle of 90 degrees

2. .... an angle larger than 180 degrees but less than 360 degrees

3. .... an angle less than 90 degrees

4. .... an angle of 180 degrees

5. .... an angle more than 90 degrees but less than 180 degrees

LESSON TWO: Computing the Perimeter of a Triangle

Instruction

A **triangle** is a figure with three sides and three angles. To find the **perimeter** of a triangle, add the lengths of the three sides. The formula is $P = a + b + c$. The lengths of the sides must be in the same units.

Example

Find the perimeter of this triangle.

4 ft.   5 ft.   3 ft.

$P = a + b + c$
$P = 5 + 4 + 3$
$P = 12$ ft.

Exercise A   Find the perimeters of these triangles.

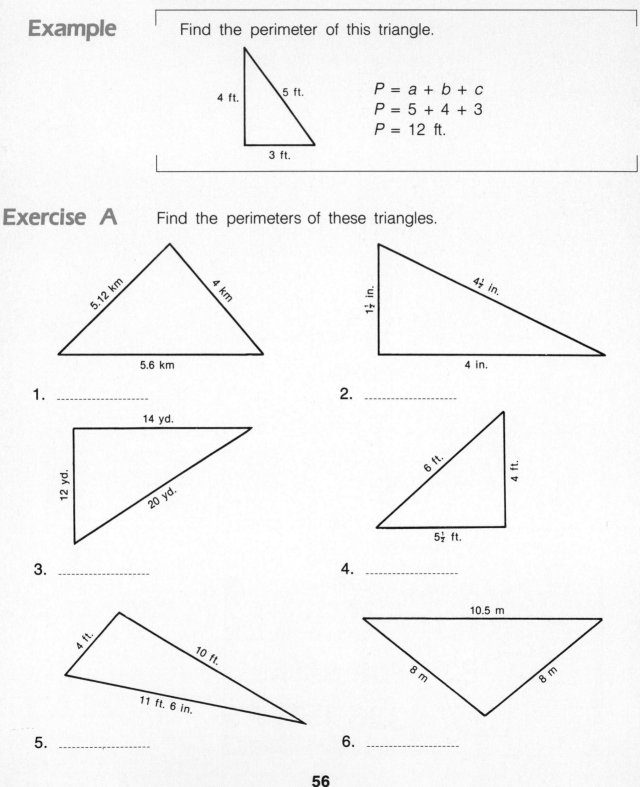

5.12 km   4 km   5.6 km

1. _____

$1\frac{1}{2}$ in. $4\frac{1}{2}$ in.   4 in.

2. _____

14 yd.   12 yd.   20 yd.

3. _____

6 ft.   4 ft. $5\frac{1}{2}$ ft.

4. _____

4 ft.   10 ft.   11 ft. 6 in.

5. _____

10.5 m   8 m   8 m

6. _____

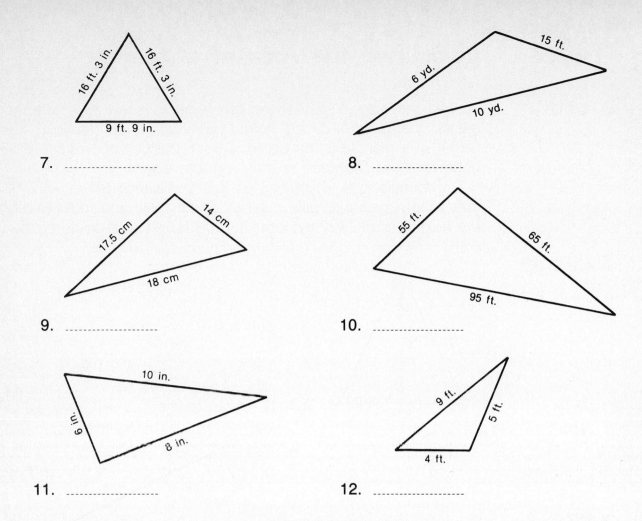

7. ....

8. ....

9. ....

10. ....

11. ....

12. ....

Exercise B   Solve the following problems.

1. Find the perimeter of a triangle having two sides of 12.3 centimeters and a base of 5.7 centimeters.   ....

2. Two sides of a triangle measure 2.5 miles, and its base is 1.5 miles. What is its perimeter?   ....

3. A very small triangle has two sides $\frac{3}{4}$ inch long and a base $\frac{5}{8}$ inch long. What is its perimeter?   ....

4. A triangular piece of property has two sides of 196 feet 4 inches and a base of 100 feet 2 inches. How much fence will be required to enclose it?   ....

5. The gable of a house forms a triangle that measures 36 feet 6 inches by 36 feet 6 inches by 74 feet 4 inches. What is its perimeter?   ....

57

LESSON THREE: Computing the Area of a Triangle

Instruction

To find the **area** of a **triangle** you need to know two measurements, the **base** and the **height** (altitude) of the triangle. The height of a triangle is the distance from one corner (vertex) to an opposite side. The base is that opposite side.

The formula is $A = \frac{1}{2}bh$, or, as it is sometimes written, $A = \frac{bh}{2}$. This simply means to take one-half of the base and multiply it by the height, or multiply the base by the height and divide by two.

Example

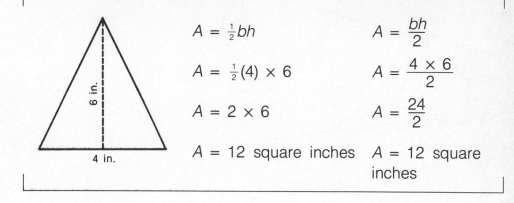

$A = \frac{1}{2}bh$

$A = \frac{1}{2}(4) \times 6$

$A = 2 \times 6$

$A = 12$ square inches

$A = \frac{bh}{2}$

$A = \frac{4 \times 6}{2}$

$A = \frac{24}{2}$

$A = 12$ square inches

Exercise A   Solve the following problems.

1. The gable of a house forms a triangle. Hernando painted the gable of his house. The base is 36 feet 6 inches and the altitude is 12 feet. Find the area of the surface Hernando painted.   ....

2. A triangle has a base of 216.4 centimeters and an altitude of 93.6 centimeters. What is its area?   ....

3. Find the area of a triangle with a base of 4 miles and an altitude of 7 miles.   ....

4. Lucia built an A-frame camp house. The ends form two triangles, each with a base of 26 feet and a height of 16 feet. What is the total area of both ends?   ....

5. A triangular-shaped park whose base is 120 feet and whose altitude is 90 feet is going to be sodded at a cost of 53¢ per square foot. How much is it going to cost?   ....

58

6. What is the area of a triangle with a base of 83 meters and an altitude of 96 meters? ....

7. A pyramid is formed of four equal triangles. If each has a base of 186 feet and a height of 103 feet, what is the total surface area of the pyramid? ....

8. The figure below is made up of two equal triangles. Each has a base of 28 feet and a height of 32 feet. What is the total area of the figure? ....

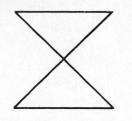

9. What is the area of this triangle? ....

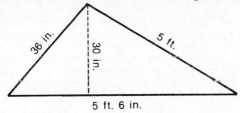

Exercise B   Solve the following problems.

1. Find the areas of these triangles.

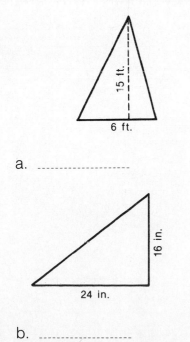

   a. ....

   b. ....

59

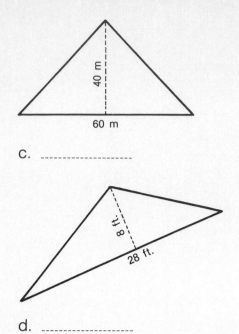

40 m

60 m

c. ------------------

8 ft.

28 ft.

d. ------------------

2. Find the area of a triangle having the same base as triangle c above but one-half the altitude.

----

3. A real estate company uses triangular-shaped signs to advertise houses for sale. Each sign is 2 feet high and has a base of 3 feet. How much area will have to be painted for 24 of these signs?

----

4. Amanda has a flower bed in the shape of a triangle. The height is 9 feet, and the base is 6 feet. She wishes to plant pansies in the bed. How many square feet are to be planted?

----

5. The Holdens built a triangular deck. The base of the triangle is 21 feet and the height is 12 feet. What is the area of the deck?

----

6. How many square yards of fabric are required to make a triangular sail for a boat if the base must be 8 feet and the height 18 feet? (1 square yard contains 9 square feet.)

----

7. The Garzas' vegetable garden in their back-yard is triangular shaped. The base is 20 feet; the height is 15 feet. What is the garden's area?

UNIT 4—REVIEW

Exercise A   Answer the following questions.

1. Define these types of angles.

   a. acute _____

   b. right _____

   c. obtuse _____

   d. straight _____

   e. reflex _____

2. What is the formula for finding the perimeter of a triangle? _____

3. Find the perimeters of these triangles.

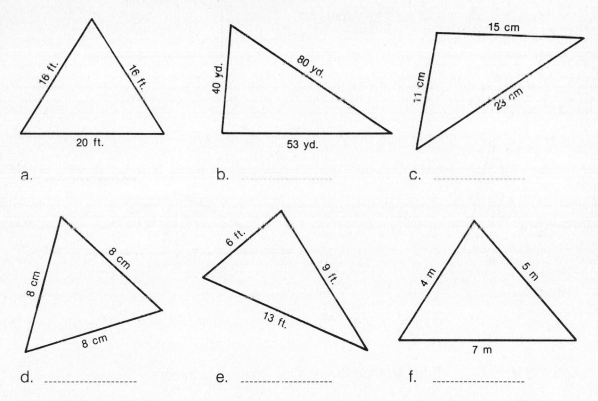

a.

b. _____

c. _____

d. _____

e. _____

f. _____

4. Find the areas of these triangles.

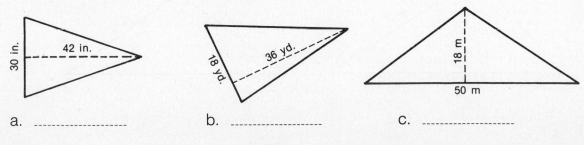

a. _____

b. _____

c. _____

61

Exercise B

Match the types of angles to their descriptions. Write the correct letter in each blank.

1. _____ reflex
2. _____ obtuse
3. _____ acute
4. _____ right
5. _____ straight

a. less than 90 degrees

b. an angle of 90 degrees

c. more than 90 degrees but less than 180 degrees

d. an angle of 180 degrees

e. larger than 180 degrees but less than 360 degrees

Exercise C

Find the areas of these triangles.

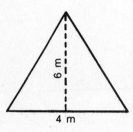

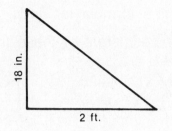

1. _____

2. _____

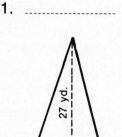

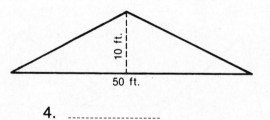

3. _____

4. _____

Exercise D

Find the perimeters of these triangles.

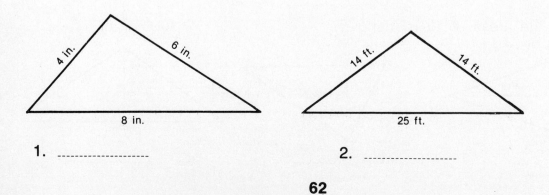

1. _____

2. _____

FINDING AREAS OF IRREGULAR-SHAPED ROOMS

When working with everyday items, you will find that many of the shapes you will be dealing with will be irregular. However, most irregular shapes can be divided into two or more regular shapes, such as squares, triangles, and rectangles. By dividing an irregular shape into regular shapes, you will be able to easily find its area. Look at the shape below. It has been divided into several regular shapes with dotted lines. By finding the areas of each of the five shapes and then adding the areas, you could find the total area of the entire shape.

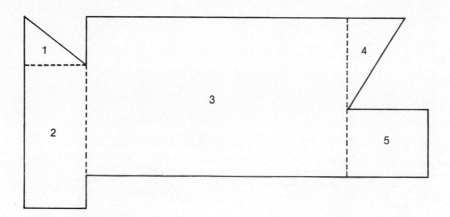

Directions Nicholas and Erin Bournias want to put carpet in their living room and dining room. To get an idea of how much it will cost, they need to find out how much carpet they will need. Look at the floor plan below and calculate how many square yards of carpet they need. (1 square yard = 9 square feet.) _____

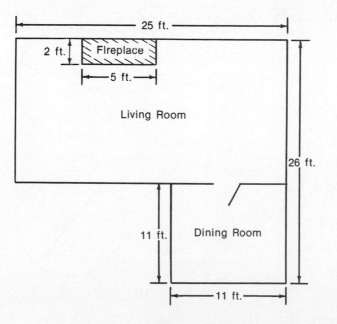

UNIT 5—PRETEST
CIRCLES AND CYLINDERS

Find the circumference of each circle. Use $\frac{22}{7}$ for π.

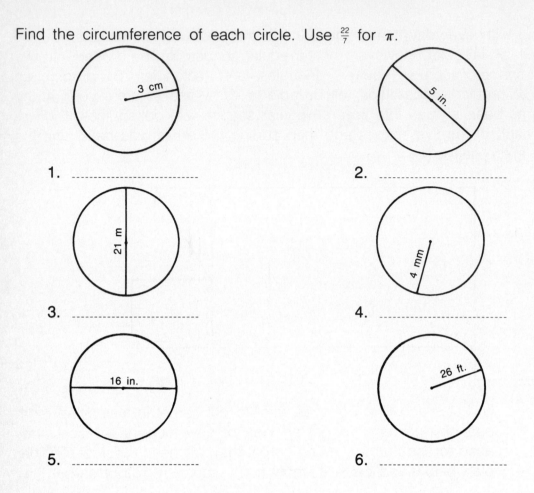

1. _____

2. _____

3. _____

4. _____

5. _____

6. _____

Find the area of each circle. Use 3.14 for π.

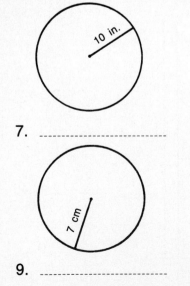
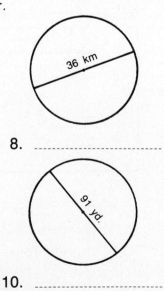

7. _____

8. _____

9. _____

10. _____

Find the volumes of these cylinders.

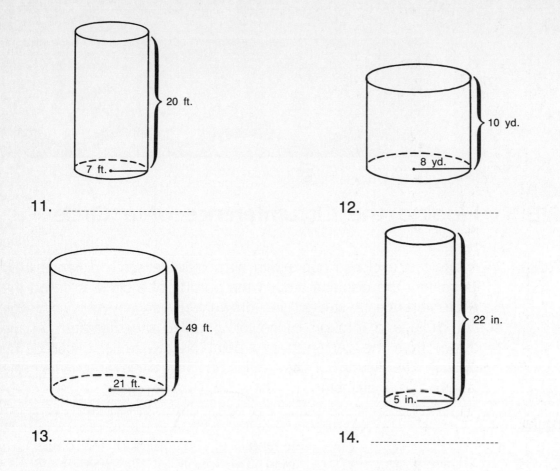

11. _____

12. _____

13. _____

14. _____

Solve the following problems, using 3.14 for π. Blacken the letter to the right that corresponds to the letter before each correct answer.

15. A caterer's largest pot has a radius of 11 inches. It is 26 inches high. What is the capacity of the pot?
a. 9,678.44 cubic in.   b. 9,878.44 cubic in.   c. 98.7844 cubic in.   d. 98,784.4 cubic in.

[a] [b] [c] [d]

16. Find the volume of a cylinder which has a height of 18 inches and a radius of 2 inches.
a. 22.60 cubic in.   b. 2.26 cubic in.   c. 226.08 cubic in.   d. 228.06 cubic in.

[a] [b] [c] [d]

17. How many cubic yards of dirt will be needed to fill a circular hole 15 feet deep if the hole has a diameter of 14 feet?
a. 2,307.9 cubic ft.   b. 230.79 cubic ft.   c. 23.07 cubic ft.   d. 2,203.9 cubic ft.

[a] [b] [c] [d]

18. There are 231 cubic inches in a gallon. A plastic container has a diameter of 8 inches and a height of 7 inches. Will it hold a gallon of potato salad?
a. yes   b. no

[a] [b]

CIRCLES AND CYLINDERS

LESSON ONE:

Finding the Circumference of a Circle

Instruction

A **circle** is a closed curve with all points the same distance from its center. The distance around the outside of a circle is called the **circumference**. A straight line drawn between any two points on the circle and through its center is called the **diameter**. A line drawn from the center to any point on the circle is called the **radius**. The length of the radius of any circle is one-half the length of the diameter.

Example

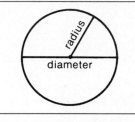

The length of the radius of this circle is $\frac{1}{2}$ inch. The length of the diameter is 1 inch.

The formula for finding the circumference of a circle is $C = \pi d$. It means the circumference is equal to the value of π times the length of the diameter of the circle. The symbol π (called *pi*) is about $3\frac{1}{7}$ ($\frac{22}{7}$) or 3.1416. Either the fraction or decimal may be used when figuring the circumference of a circle.

Example

Find the circumference of a circle that has a diameter of 12 feet.

$C = \pi d$

$C = \frac{22}{7} \times \frac{12}{1}$

$C = \frac{264}{7}$

$C = 37\frac{5}{7}$ feet

$C = \pi d$

$$
\begin{array}{r}
C = 3.1\,4\,1\,6 \\
\times\ 1\,2 \\
\hline
6\,2\,8\,3\,2 \\
3\,1\,4\,1\,6 \\
\hline
3\,7.6\,9\,9\,2 \text{ feet}
\end{array}
$$

12 ft.

Since $\frac{22}{7}$ and 3.1416 are approximate values of π, the answers are about the same but not exactly the same.

Remember, when you know the length of the radius of a circle, you must multiply it by 2 to find the diameter.

Example

What is the circumference of a circle whose radius is 7 centimeters?

radius = 7 cm
diameter = 7 cm × 2 = 14 cm

$$C = \pi d$$

$$C = \frac{22}{\underset{1}{7}} \times \frac{\overset{2}{14}}{1}$$

$$C = 44 \text{ centimeters}$$

Exercise A

Solve the following problems. Use $\frac{22}{7}$ for π.

1. If a circle has a radius of 56 feet, what is its circumference? ------------------

2. Find the difference in the circumferences of two circles if the second one is twice as large as the first. The first circle has a diameter of 49 meters. ------------------

3. What is the circumference of a circle with a radius of $\frac{1}{4}$ mile? ------------------

4. Edna has a plastic swimming pool that is round. Its diameter is 5 feet. What is its circumference? ------------------

5. Leo wanted to make a rose garden in the shape of a circle. He drove down a stake at the center and tied a rope to it. From the stake to the end of the rope measured 6 feet. What will be the circumference of the rose garden? ------------------

6. Melba wishes to place a lace border around a round tablecloth. The tablecloth is 42 inches across. How much lace will she need to go around it? ------------------

7. The diameter of a merry-go-round is 63 feet. Each person gets to make 23 rounds per ride. How far does each person ride? ------------------

8. A large round advertising balloon is 15 feet in diameter. What is its circumference? ------------------

9. Wilma has a train set. The track is circular. The diameter is 7 feet. How many trips around the track must the train go to travel 1 mile (5,280 ft.)?

....

10. The radius of a water tank is 8 feet. What is the distance around the tank?

....

11. A zoo has a circular pool for its polar bears. How much fence is needed to enclose the pool if its diameter is 28 feet?

....

Exercise B

Complete the table. Then find the circumference of each circle. Use $\frac{22}{7}$ for π.

	Radius	Diameter	Circumference
1.		56	
2.	3.5		
3.	6		
4.		140	
5.		392	

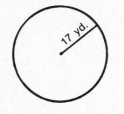

6. ....

7. ....

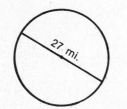

8. ....

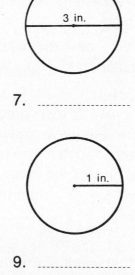

9. ....

10. ....

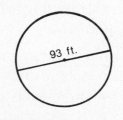

11. ....

LESSON TWO: Finding the Area of the Circle

Instruction

To find the **area** of a **circle** use this formula: $A = \pi r^2$. r^2 is read "r squared" and means r × r. Note that the area of a circle is found by using the radius. When you square a number, remember that it is multiplied, not added.

Example

What is the area of a circle whose radius is 14 feet?

$A = \pi r^2$

$A = \frac{22}{7} \times 14 \times 14$

$A = \frac{22}{\cancel{7}_1} \times \frac{\cancel{14}^2}{1} \times \frac{14}{1}$

$A = 616$ square feet

If you know the diameter of a circle, be sure to divide it by 2 ($r = \frac{1}{2}d$) before finding the area.

Example

Find the area of the circle whose diameter is 56 feet.

$r = \frac{1}{2}d$ $A = \pi r^2$

$r = \frac{56}{2}$ ⟶ $A = \frac{22}{7} \times 28 \times 28$

$r = 28$ feet $A = \frac{22}{\cancel{7}_1} \times \frac{\cancel{28}^4}{1} \times \frac{28}{1}$

$A = 2{,}464$ square feet

Exercise A

Complete the table. Then find the areas of the circles. Use $\frac{22}{7}$ for π.

	Radius	Diameter	Area
1.	42		
2.		14	
3.	$9\frac{1}{3}$		
4.	$5\frac{1}{4}$		
5.		28	

Courtesy Brown & Sharpe Manufacturing Company

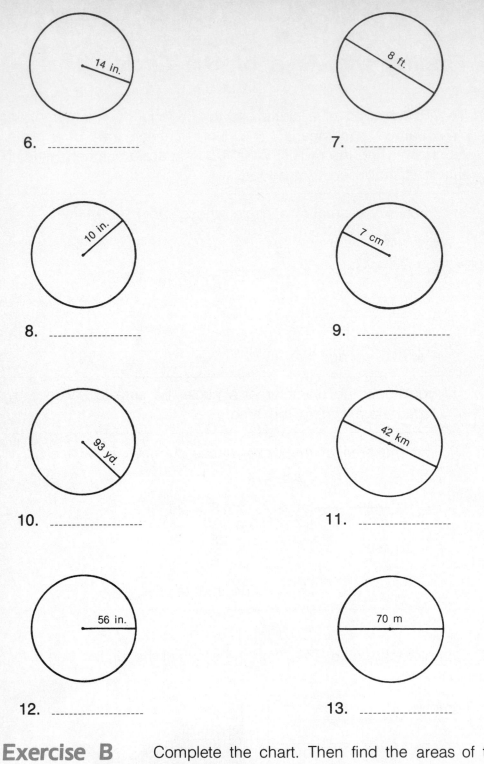

6. _____

7. _____

8. _____

9. _____

10. _____

11. _____

12. _____

13. _____

Exercise B

Complete the chart. Then find the areas of the circles. Use $\frac{22}{7}$ for π for problems 6-11.

	Diameter	Radius	Use for π	Area
1.		9	3.14	
2.		100	3.14	
3.	10		$\frac{22}{7}$	
4.	21		$\frac{22}{7}$	
5.		15	3.14	

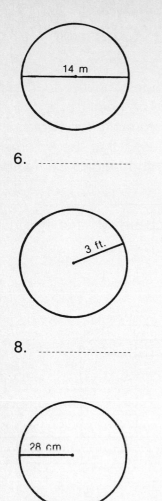

6. _____

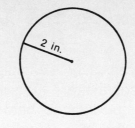

7. _____

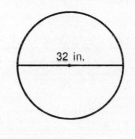

8. _____

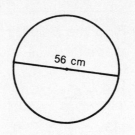

9. _____

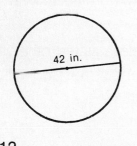

10. _____

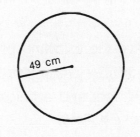

11. _____

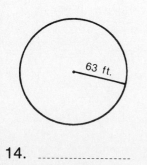

12. _____

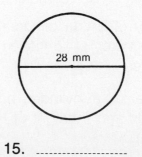

13. _____

14. _____

15. _____

LESSON THREE: Finding the Volume of the Cylinder

Instruction

A can is an example of a **cylinder**. Cylinders have two circular bases that are the same size and parallel. The height of a cylinder is the distance between the bases. The formula for finding the **volume** of a **cylinder** is $V = \pi r^2 h$. This means that the volume is equal to π times the radius squared times the height. Remember, volume is expressed in cubic units.

Example

Find the volume of a can that is 6 inches high and has a diameter of 4 inches.

$$r = \frac{1}{2}d$$

$$r = \frac{1}{2} \times \frac{4}{1} = 2$$

$$V = \pi r^2 h$$

$$V = \frac{22}{7} \times (2)^2 \times 6$$

$$V = \frac{22}{7} \times 4 \times 6$$

$$V = \frac{22}{7} \times \frac{24}{1}$$

$$V = \frac{528}{7}$$

$$V = 75\frac{3}{7} \text{ cubic inches}$$

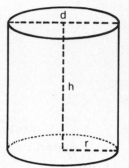

Exercise A   Solve the following problems. Use $\frac{22}{7}$ for π unless stated otherwise.

1. How many gallons will a cylinder hold if it has a radius of $3\frac{1}{2}$ feet and a height of 12 feet? (1 cu. ft. = 7.5 gal.)   ------------------

2. What is the volume of a cylindrical storage bin with an inside diameter of 14 meters and an inside height of 10 meters?   ------------------

3. Find the capacity of a can of tomato juice which has a diameter of 2 inches and a height of 7 inches.   ------------------

4. Find the volume of a cylinder having a radius of 10 yards and a height of 8 yards. (Use 3.14 for π.)   ------------------

5. Jacqueline dug a circular goldfish pond having a diameter of 28 feet. How many gallons of water are needed to fill the pond to a depth of 2 feet? (1 cu. ft. = 7.5 gallons.)　......................

6. A container stands 2 feet high and has a diameter of 14 inches. What is its volume?　......................

7. Mr. Nunez bought a slow cooker. It is 15 inches tall and has an inside diameter of 10 inches. Find the volume of the pot.　......................

8. The Barstows dug a hole for a septic tank. It measured 5 feet across and 8 feet deep. How many feet of dirt did they remove?　......................

9. A certain can is round. Its radius is 7 inches. It is 6 inches high. Will it hold 5 gallons of water? (1 gal. = 231 cu. in.)　......................

Exercise B

Solve the following problems. Then complete the chart. Use $\frac{22}{7}$ for π unless stated otherwise.

1. Naome catches rainwater in a barrel. She uses it for watering plants. How many gallons of water will she have if the barrel has an inside diameter of 3 feet and contains 4 feet of water? (1 cu. ft. = 7.5 gal.)　......................

2. A can in the shape of a cylinder is 4 inches wide and 8 inches high. What is the volume of the can in cubic inches? (Use 3.14 for π.)　......................

3. Find the volume of a cylinder which has a height of 14 inches and a radius of 2 inches. (Use 3.14 for π.)　......................

4. If the height in problem 3 is half as much, how much less is the capacity of the cylinder?　......................

5. What is the volume of a cylinder with a radius of $4\frac{2}{3}$ feet and a height of 6 feet? ....

6. How many cubic yards of dirt will be needed to fill a circular hole 10 yards deep if the hole has a diameter of 14 yards? ....

	Radius	Diameter	Height	Volume
7.	5		25	
8.	4		8	
9.	$5\frac{1}{4}$		$15\frac{3}{4}$	
10.		18	18	
11.		56	123	
12.	21		3	
13.	7		12	
14.	10		20	
15.	$4\frac{1}{2}$		9	
16.	6		15	
17.		9	11	
18.		7	12	
19.		8	10	
20.		20	30	
21.		18	13	
22.	12		7	
23.		34	20	
24.		16	16	
25.	9		22	

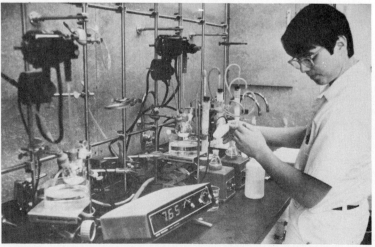

Courtesy Pfizer, Inc.

74

UNIT 5—REVIEW

Exercise A Find the circumference and area of each circle. Use $\frac{22}{7}$ for π.

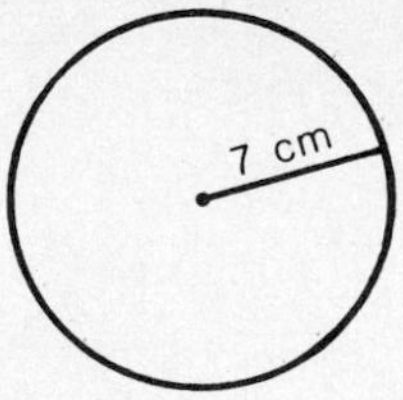

1. circumference = _____

   area = _____

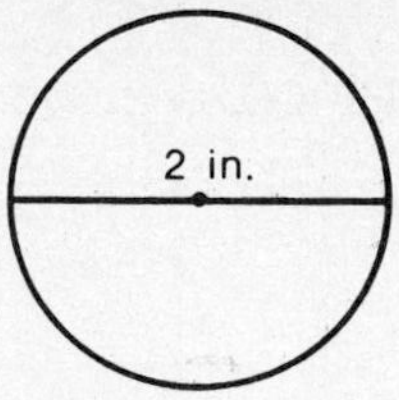

2. circumference = _____

   area = _____

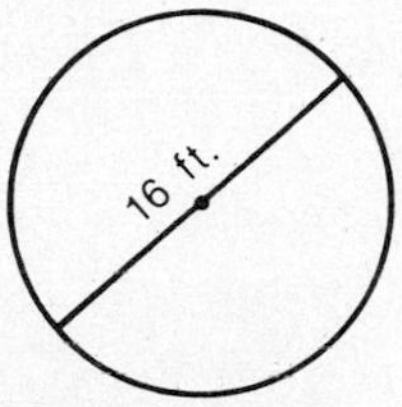

3. circumference = _____

   area = _____

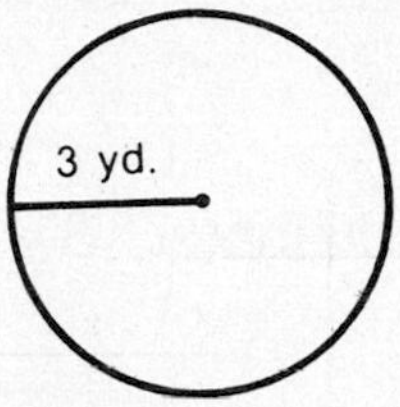

4. circumference = _____

   area = _____

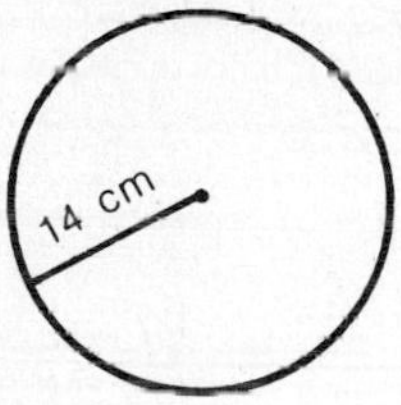

5. circumference = _____

   area = _____

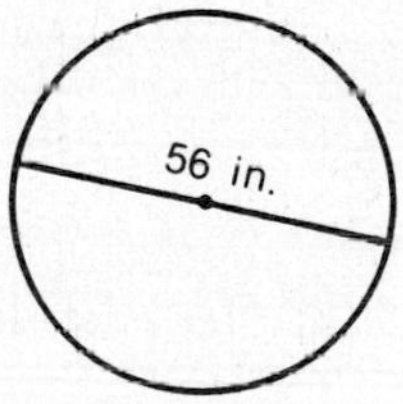

6. circumference = _____

   area = _____

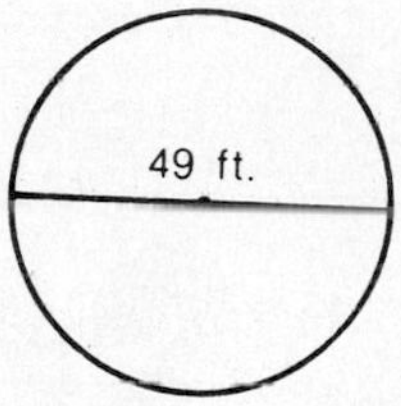

7. circumference = _____

   area = _____

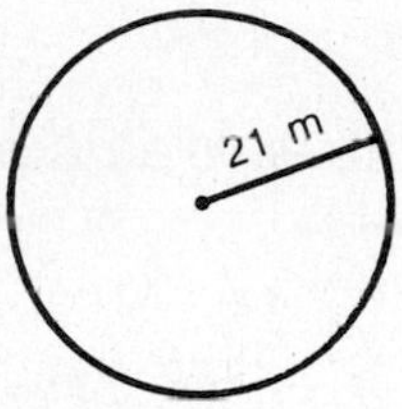

8. circumference = _____

   area = _____

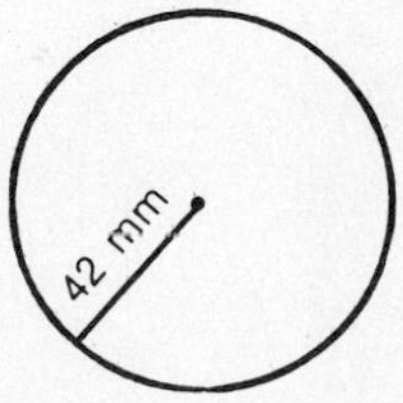

9. circumference = _____

   area = _____

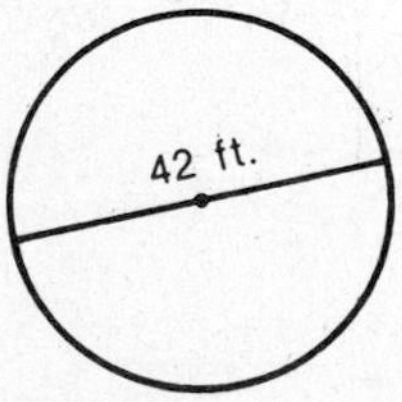

10. circumference = _____

    area = _____

75

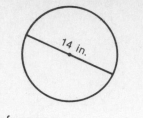

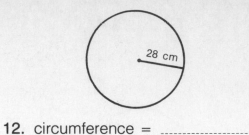

11. circumference = ...

      area = ...

12. circumference = ...

      area = ...

Exercise B

Find the volumes of cylinders with the following dimensions. Use $\frac{22}{7}$ for π.

1. 7-foot radius, 6-foot height

----

2. 16-yard diameter, 3-yard height

----

3. 14-meter radius, 12-meter height

----

4. 4-inch diameter, 8-inch height

----

5. 14-foot diameter, 35-foot height

----

6. 21-inch radius, 12-inch height

----

7. 28-centimeter diameter, 20-centimeter height

----

8. 6-meter diameter, 7-meter height

----

9. 28-inch diameter, 35-inch height

76

Exercise C

Solve the following problems. Use $\frac{22}{7}$ for π. Blacken the letter to the right that corresponds to the letter before each correct answer.

1. What is the area of a circle with a radius of 7 meters?
   a. 152 sq. meters   b. 154 sq. meters   c. 15.4 sq. meters   d. 1.54 sq. meters

   a b c d

2. A tank with a 14-foot diameter is 20 feet tall. How many gallons of oil will it hold? (1 cu. ft. = 7.5 gal.)
   a. 2300 gallons   b. 23,600 gallons   c. 23,100 gallons   d. 231 gallons

   a b c d

3. What is the circumference of a circle whose radius is 2 inches?
   a. $12\frac{4}{7}$ in.   b. $12\frac{1}{2}$ in.   c. 16 in.   d. $12\frac{2}{7}$ in.

   a b c d

4. There are 231 cubic inches in a gallon. Sarah has a container with a diameter of 7 inches and a height of 6 inches. Will it hold a gallon?
   a. yes   b. no

   a b

5. A circular flower garden has a radius of 16 feet. You want to lay a circular sidewalk 2 feet wide around the entire garden. If the walk will be 3 inches thick, how many yards of concrete will you need to order from AAA Concrete Company?
   a. 3 cubic yds.   b. $2\frac{1}{2}$ cubic yds.   c. 6 cubic yds.
   d. 2 cubic yds.

   a b c d

6. Find the areas of these circles.

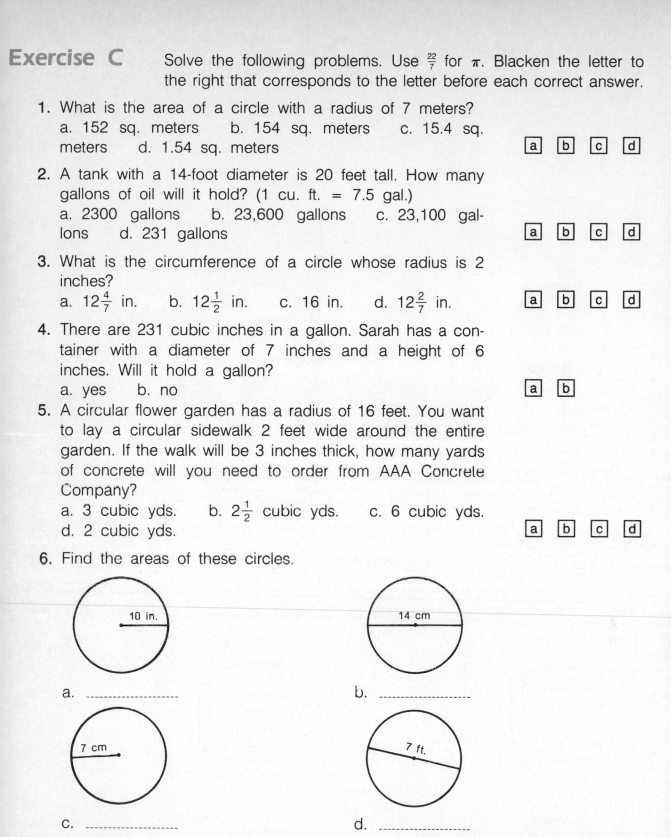

a. _____   b. _____

c. _____   d. _____

e. _____   f. _____

7. Find the circumferences of these circles.

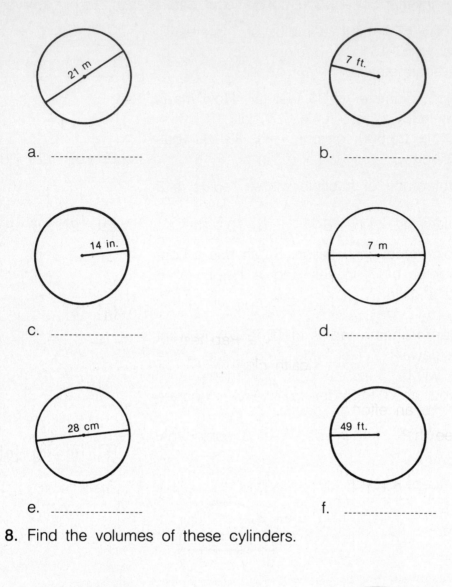

a. ....

b. ....

c. ....

d. ....

e. ....

f. ....

8. Find the volumes of these cylinders.

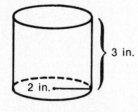

3 in.

2 in.

a. ....

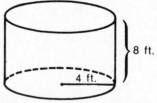

8 ft.

4 ft.

b. ....

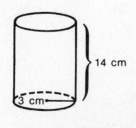

14 cm

3 cm

c. ....

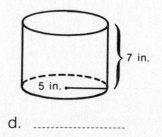

7 in.

5 in.

d. ....

HEALTH CHECKLIST

Many people take their health for granted. Most people want to be strong and healthy, but many do not take good care of themselves. Some people do not take good care of their health because they say:
- they do not have time
- they do not want to lose any pay by missing work
- they cannot afford to pay for treatment of illnesses or for checkups
- they feel pretty good most of the time

No one can afford to do nothing about his or her health or the health of family members. You should not neglect your or your family's health for any reason. Below is a health checklist. Answer each question by putting a check in the *yes* or *no* box. Were you able to mark the *yes* box beside each question?

Yes	No	
☐	☐	Do you get a chest X-ray every year?
☐	☐	If you are a woman, do you have a Pap test at least once a year?
☐	☐	Do you go to a doctor or health clinic when you are worried about your health?
☐	☐	Do you make an effort to have a checkup regularly?
☐	☐	Do you see that your children have checkups regularly?
☐	☐	When someone in your family is sick, do you take his or her temperature?
☐	☐	Have all members of your family been vaccinated against preventable diseases?
☐	☐	Do you know when to keep your children home from school because of illness?
☐	☐	Do you know the seven warning signals of cancer?
☐	☐	Do you know how to help protect yourself from heart disease?
☐	☐	Do you know when a doctor is needed when someone is sick?
☐	☐	Do you know most of the common signs of illness?

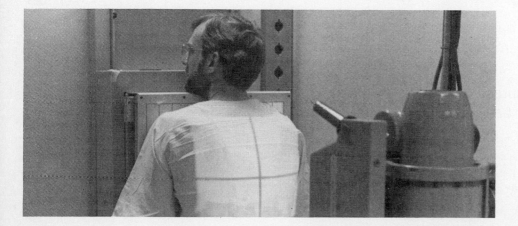

FINAL REVIEW

Solve the following problems. Blacken the letter to the right that corresponds to the letter before each correct answer.

1.   3 lb. 3 oz.   a. 5 lb. 17 oz.   b. 6 lb. 1 oz.   [a] [b] [c]
   + 2 lb. 14 oz.   c. 1 lb. 9 oz.

2.   9 gal. 2 qt.   a. 5 gal. 1 qt.   b. 4 gal. 3 qt.   [a] [b] [c]
   − 4 gal. 3 qt.   c. 13 gal. 5 qt.

3. 6 bu. 2 pk.   a. 52 bu.   b. 48 bu. 16 pk.   [a] [b] [c]
      × 8   c. 6 bu. 16 pk.

4. 12 pk. 4 qt. ÷ 4 =   a. 4 pk.   b. 1 pk. 1 qt.   [a] [b] [c]
                          c. 3 pk. 1 qt.

Find the larger unit in each pair. Blacken the letter to the right that corresponds to the letter before each correct answer.

5. a. 1 centimeter   b. 5 millimeters   [a] [b]
6. a. 1 kilometer   b. 100 milliliters   [a] [b]
7. a. 100 meters   b. 100 centimeters   [a] [b]
8. a. 8 grams   b. 40 milligrams   [a] [b]

Blacken the letter to the right that corresponds to the letter before each correct answer.

9. What is the basic unit for measuring mass in the metric system?
   a. liter   b. meter   c. gram   [a] [b] [c]

10. The basic metric unit of length is a
   a. liter   b. gram   c. meter   [a] [b] [c]

11. At what temperature does water freeze, according to the Celsius thermometer?
   a. 360°   b. 0°   c. 100°   [a] [b] [c]

12. What is the basic metric unit for measuring capacity?
   a. liter   b. meter   c. gram   [a] [b] [c]

Convert these measures. Blacken the letter to the right that corresponds to the letter before each correct answer.

13. 8 ounces ≈ ══ grams
   a. 64   b. 152   c. 224   [a] [b] [c]

14. 112 grams ≈ ══ ounces
   a. 7   b. 4   c. 8   [a] [b] [c]

15. 4 pounds 4 ounces ≈ ══ grams
    a. 1,912   b. 1,206   c. 16   `a` `b` `c`

16. 8.8 pounds ≈ ══ kilograms
    a. 4   b. 8   c. 2   `a` `b` `c`

Answer the following questions. Black the letter to the right that corresponds to the letter before each correct answer.

17. Would you be comfortable swimming in 100°C water?
    a. yes   b. no   `a` `b`

18. Does body temperature of 98.6°C indicate fever?
    a. yes   b. no   `a` `b`

19. Which is colder?
    a. −10°F   b. −10°C   `a` `b`

20. Would a person need heavy clothing during 30°C temperature?
    a. yes   b. no   `a` `b`

Compute the perimeters of these squares. Blacken the letter to the right that corresponds to the letter before each correct answer.

21. s = 71 inches
    a. 142 inches   b. 5,041 inches   c. 284 inches   `a` `b` `c`

22. s = 6.6 meters
    a. 39.6 meters   b. 13.2 meters   c. 26.4 meters   `a` `b` `c`

23.

$7\frac{1}{4}$ yards

    a. 29 yards   b. $14\frac{1}{2}$ yards   c. 21.75 yards   `a` `b` `c`

24.

6.2 km

    a. 49.6 km   b. 24.8 km   c. 37.2 km   `a` `b` `c`

Compute the areas of these squares. Blacken the letter to the right that corresponds to the letter before each correct answer.

25. s = .2 inches
    a. .4 square inches   b. 4 square inches
    c. .04 square inches   `a` `b` `c`

81

26. s = 220 feet

   a. 850 square feet   b. 440 square feet

   c. 48,400 square feet.   a b c

27.

33.1 m

   a. 1,095.61 square meters   b. 66.2 square meters

   c. 132.4 square meters   a b c

28.

12 cm

   a. 1,144 square centimeters b. 1,044 square centimeters

   c. 144 square centimeters   a b c

Compute the total surface areas of these cubes. Blacken the letter to the right that corresponds to the letter before each correct answer.

29. s = 8 centimeters

   a. 384 square centimeters   b. 64 square centimeters

   c. 16 square centimeters   a b c

30. s = 3.7 meters

   a. 821.4 square meters   b. 82.14 square meters

   c. 13.69 square meters   a b c

31.

6.2 ft.   6.2 ft.   6.2 ft.

   a. 238.328 square feet   b. 13.6 square feet

   c. 230.64 square feet   a b c

32.

6 in.   6 in.   6 in.

   a. 216 square inches   b. 18 square inches

   c. 108 square inches   a b c

Find the perimeters of these rectangles. Blacken the letter to the right that corresponds to the letter before each correct answer.

33. *l* = 20 feet; *w* = 32 feet
    a. 52 feet   b. 104 feet   c. 520 feet   [a] [b] [c]

34. *l* = 2.6 meters; *w* = 4.8 meters
    a. 14.8 meters   b. 7.4 meters   c. 12.48 meters   [a] [b] [c]

35.
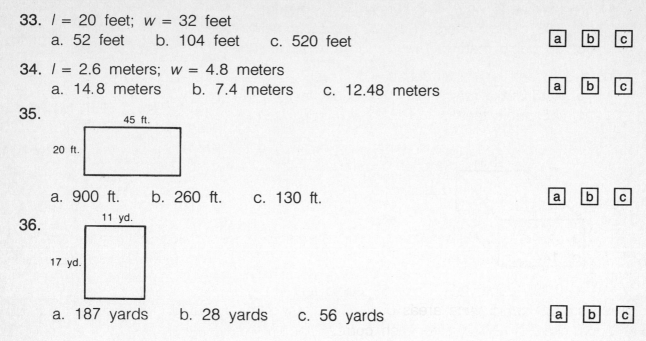

    a. 900 ft.   b. 260 ft.   c. 130 ft.   [a] [b] [c]

36.

    a. 187 yards   b. 28 yards   c. 56 yards   [a] [b] [c]

Find the areas of these rectangles. Blacken the letter to the right that corresponds to the letter before each correct answer.

37. *l* = 2.9 feet; *w* = 9.2 feet
    a. 266.8 square feet   b. 26.68 square feet
    c. 2,668 square feet   [a] [b] [c]

38. *l* = 17 rods; *w* = 17.5 rods
    a. 34.5 rods   b. 297.5 rods   c. 297.5 square rods   [a] [b] [c]

39.
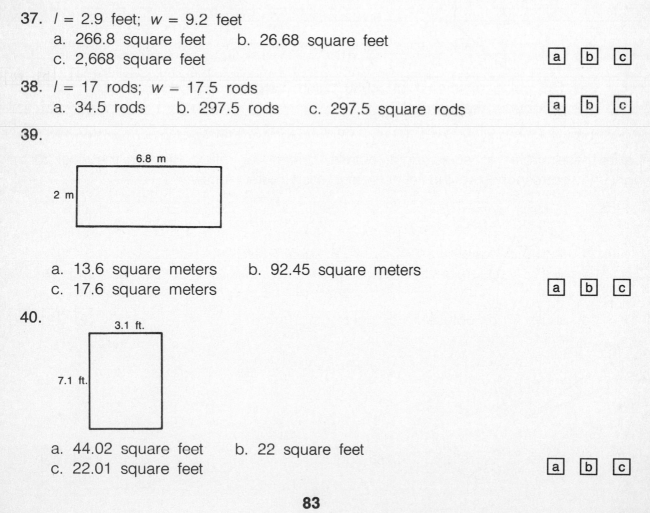

    a. 13.6 square meters   b. 92.45 square meters
    c. 17.6 square meters   [a] [b] [c]

40.

    a. 44.02 square feet   b. 22 square feet
    c. 22.01 square feet   [a] [b] [c]

Find the volumes of these rectangles. Blacken the letter to the right that corresponds to the letter before each correct answer.

41. l = 6 meters; w = 4 meters; h = 12 meters
    a. 288 cubic meters   b. 22 square meters
    c. 36 cubic meters   ⓐ ⓑ ⓒ

42. l = 3 feet; w = 1.2 feet; h = $1\frac{1}{4}$ feet
    a. 4.45 cubic feet   b. 3.75 cubic feet
    c. 4.5 cubic feet   ⓐ ⓑ ⓒ

43.

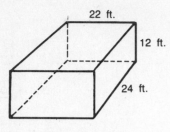

    a. 6,336 cubic feet   b. 58 cubic feet
    c. 528 cubic feet   ⓐ ⓑ ⓒ

44.

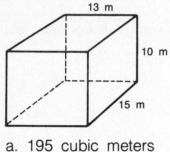

    a. 195 cubic meters   b. 1,950 cubic meters
    c. 130 cubic meters   ⓐ ⓑ ⓒ

Label these angles as acute, right, obtuse, straight, or reflex. Blacken the letter to the right that corresponds to the letter before each correct answer.

45.

    a. acute   b. obtuse   c. right   ⓐ ⓑ ⓒ

46.

    a. acute   b. reflex   c. straight   ⓐ ⓑ ⓒ

47.

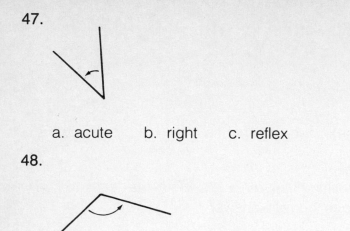

a. acute   b. right   c. reflex

48.

a. reflex   b. acute   c. obtuse

Find the areas of these triangles. Blacken the letter to the right that corresponds to the letter before each correct answer.

49.

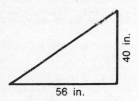

a. 22 square meters   b. 120 square meters
c. 60 square meters

50.

56 in. 40 in.

a. 2,240 square inches   b. 1,120 square inches
c. 96 square inches

Find the perimeters of these triangles. Blacken the letter to the right that corresponds to the letter before each correct answer.

51.

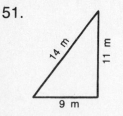

a. 34 m   b. 1,386 m   c. 340 square meters

52.

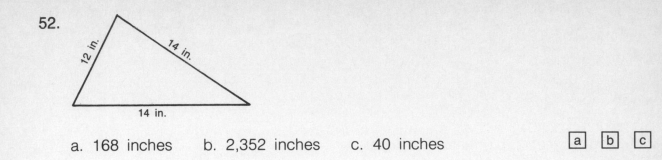

     a. 168 inches   b. 2,352 inches   c. 40 inches   a | b | c

Find the circumference of each circle, using 3.14 for π. Blacken the letter to the right that corresponds to the letter before each correct answer.

53.

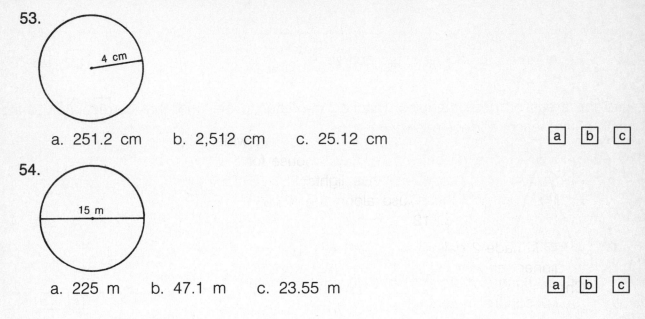

     a. 251.2 cm   b. 2,512 cm   c. 25.12 cm   a | b | c

54.

     a. 225 m   b. 47.1 m   c. 23.55 m   a | b | c

Find the volumes of these cylinders. Use 3.14 for π. Blacken the letter to the right that corresponds to the letter before each correct answer.

55. $r = 4$ ft; $h = 10$ ft.
    a. 160 cubic feet   b. 1,502.4 cubic feet
    c. 502.4 cubic feet   a | b | c

56. $d = 14$ m; $h = 100$ m
    a. 15,386 cubic meters   b. 1,400 cubic meters
    c. 750 cubic meters   a | b | c

Read and solve the following problems. Blacken the letter to the right that corresponds to the letter before each correct answer.

57. Fred is going to panel his living room. The room is 12 feet wide and 16 feet long. Each sheet of paneling is 4 feet by 8 feet. How many sheets of paneling does Fred need to buy?
    a. 14 sheets   b. 16 sheets   c. 13 sheets   a | b | c

58. Dave needs to repave the parking lot at his store. The paving company charges 50¢ per square foot. The lot measures 80 feet by 140 feet. How much will the repaving job cost?

a. $6,500   b. $5,500   c. $5,600

☐a ☐b ☐c

59. Dale needs to fill an old well with dirt. The well is 20 feet deep and 4 feet across. How much dirt will it take to fill the well?

a. 251.2 cubic ft.   b. 100.48 cubic ft.
c. 10,048 cubic ft.

☐a ☐b ☐c

60. Jerry plans to install solar film on the windows of his house. Two windows are 30 inches by 60 inches and one window is 60 inches by 60 inches. The film comes in rolls that contain 500 square inches. How many rolls will Jerry need to buy?

a. 12 rolls   b. 15 rolls   c. 14 rolls

☐a ☐b ☐c

61. Stanley's house is 110 feet long and 60 feet wide. Allowing five extra feet at each end of the house for the gables, how many strings of Christmas lights 25 feet long will it take to surround the house along the eaves?

a. 10   b. 8   c. 12   d. 14

☐a ☐b ☐c ☐d

62. Charles made 2 gallons of chili. His family ate $\frac{1}{4}$ of the chili at dinner, and he plans to freeze the rest for use later. Charles's largest freezer container measures 7 inches in diameter and is 6 inches high. Will it hold all the leftover chili?

a. yes   b. no

☐a ☐b

63. A piece of caramel candy measures 21 millimeters by 21 millimeters by 12 millimeters. What is the volume of the piece of candy?

a. 441 cu. mm   b. 5,292 cu. mm   c. 252 cu. mm
d. 3,024 cu. mm

☐a ☐b ☐c ☐d

64. There are 27 pieces of candy in a package whose net weight is 284 grams. What is the weight of one piece if each one measures 1 inch by 1 inch by $\frac{1}{2}$ inch?

a. 10.5 g   b. $\frac{1}{2}$ cu. in.   c. 1 lb.   d. 142 g

☐a ☐b ☐c ☐d

Page 1. **1.** 18 ft. 4 in.; **2.** 10 qt.; **3.** 3 gal. 3 qt.; **4.** 2 pk. 1 qt.; **5.** 12 lb. 4 oz.; **6.** 3 yd. 2 ft.; **7.** 1 millimeter; **8.** 9 grams; **9.** 50 milliliters; **10.** gram; **11.** liter; **12.** 100; **13.** 112; **14.** 956; **15.** 2; **16.** $\frac{1}{2}$; **17.** 27°F; **18.** 37°; **19.** 70°C; **20.** no

Page 4. **A. 1.** 18 ft. 2 in.; **2.** 26 qts.; **3.** 2 gal. 3 qt.; **4.** 1 pk. 4 qt.; **5.** 6 lb. 4 oz.; **6.** 2 yd. 2 ft.; **7.** 17 mi.; **8.** 3 c. 2 fl. oz.; **9.** 1 yd. 2 ft.; **10.** 16 ft. 4 in.; **11.** 1 lb. 10 oz.; **12.** 38 yd. 1 ft.; **13.** 1 gal. 3 qt.; **14.** 24 ft. 7 in.; **15.** 1 lb. 12 oz. **B. 1.** 30 yd.; **2.** 1 gal. 2 qt.; **3.** 13 mi.; **4.** 4 lb. 11 oz.; **5.** 40 bu. 2 pk.; **6.** 17 gal. 1 qt.; **7.** 1 ft.; **8.** 1 qt. 1 pt.; **9.** 18 ft. 9 in.; **10.** 16 lb. 14 oz.; **11.** 6 gal.; **12.** 13 ft. 4 in.

Page 6. **A. 1.** 1 decimeter; **2.** 1 dekameter; **3.** 1 millimeter; **4.** 1 centimeter; **5.** 6 dekameters; **6.** 1 meter; **7.** 2 decimeters; **8.** 2 centimeters; **9.** 5 centimeters; **10.** 1 millimeter; **11.** 1 centimeter; **12.** 2 centimeters; **13.** 10 centimeters; **14.** 1,000 millimeters; **15.** 800 millimeters; **16.** 2,000 millimeters;

Page 7. **17.** 86 meters; **18.** 200 millimeters; **19.** 100 millimeters; **20.** 1 meter. **B. 1.** 1 decimeter; **2.** 1 meter; **3.** 150 meters; **4.** 3 kilometers; **5.** 275 centimeters; **6.** 1 kilometer; **7.** 3 kilometers; **8.** 55 centimeters; **9.** 10 kilometers; **10.** 2 kilometers; **11.** 1 kilometer; **12.** 1 kilometer; **13.** 3 meters; **14.** 91 meters; **15.** 50 dekameters; **16.** 2 kilometers; **17.** 1 meter; **18.** 1 meter; **19.** 1 centimeter; **20.** 1 meter

Page 8. **A. 1.** foot, meter; **2.** inch, centimeter; **3.** mile, kilometer; **4.** inch, centimeter; **5.** yard (foot), meter; **6.** inch, centimeter; **7.** foot, meter; **8.** inch, centimeter; **9.** foot, meter; **10.** inch, centimeter; **11.** inch, centimeter; **12.** inch, centimeter; **13.** inch, centimeter; **14.** inch, centimeter; **15.** foot, meter; **16.** foot, meter; **17.** foot, meter; **18.** inch, centimeter

Page 9. **B. 1.** inch, centimeter; **2.** inch, centimeter; **3.** inch, centimeter; **4.** inch, centimeter; **5.** yard (foot), meter; **6.** foot, meter; **7.** inch, centimeter; **8.** foot, meter; **9.** inch, centimeter; **10.** foot, meter; **11.** foot (yard), meter; **12.** foot, meter; **13.** foot (yard), meter; **14.** inch, centimeter; **15.** inch, centimeter; **16.** inch, centimeter; **17.** inch, centimeter; **18.** foot, meter; **19.** inch, centimeter; **20.** inch, centimeter

Page 10. **A. 1.** 5 cm; **2.** 7 cm; **3.** 11 cm; **4.** 12 cm; **5.** 3 cm; **6.** 10 cm; **7.** 7 cm

Page 11. **B. 1.** 10 cm; **2.** 2 cm; **3.** 8 cm; **4.** 12 cm; **5.** 9 cm; **6.** 5 cm; **7.** 4 cm; **8.** 9 cm

Page 12. Answers will vary.

Page 14. **A. 1.** 1 kilogram; **2.** 1 dekagram; **3.** 1 kilogram; **4.** 1 decigram; **5.** 1 kilogram; **6.** 1 gram; **7.** 5 hectograms. **8.** 50 decigrams; **9.** 1 kilogram; **10.** 3 dekagrams. **B. 1.** p; **2.** o; **3.** h; **4.** e; **5.** i; **6.** c; **7.** a; **8.** m; **9.** g; **10.** b; **11.** l; **12.** j; **13.** f; **14.** n; **15.** k; **16.** d

Page 15. **A. 1.** 56; **2.** 98; **3.** 280; **4.** 420;

Page 16. **5.** 506; **6.** 1; **7.** 126; **8.** .5; **9.** 3; **10.** 42; **11.** 4; **12.** 112; **13.** 173.6; **14.** 30.8; **15.** 1,800; **16.** 2; **17.** 11.2; **18.** .5; **19.** 2; **20.** 1.1

Page 17. **B. 1.** 112; **2.** 280; **3.** 252; **4.** 1; **5.** 450; **6.** 534; **7.** 900; **8.** 336; **9.** 105.6; **10.** 2.5; **11.** 38.72; **12.** 2; **13.** 4; **14.** 896 **15.** 11; **16.** 3,150

Page 18. **A. 1.** 1 hectoliter; **2.** 1 milliliter; **3.** 8 liters; **4.** 15 liters; **5.** 1 dekaliter; **6.** 1 liter; **7.** 60 centiliters; **8.** 10 deciliters; **9.** 9 liters; **10.** 15 hectoliters; **11.** 100 centiliters

Page 19. **B. 1.** 1 dekaliter; **2.** 1 liter; **3.** 1 liter; **4.** 200 centiliters; **5.** 500 liters; **6.** 10 centiliters; **7.** 2 kiloliters; **8.** 59 liters; **9.** 16 deciliters; **10.** 1 dekaliter; **11.** 1 kiloliter; **12.** 1 centiliter; **13.** 2 deciliters; **14.** 75 centiliters; **15.** 250 dekaliters; **16.** 99 hectoliters; **17.** 10 dekaliters; **18.** 83 liters; **19.** 896 hectoliters; **20.** 15 liters; **21.** 2 dekaliters; **22.** 1 kiloliter; **23.** 1 dekaliter; **24.** 60 liters; **25.** 1 kiloliter

Page 20. **A. 1.** yes; **2.** yes; **3.** 25°F.; **4.** 23°C; **5.** 19°C; **6.** 14°C; **7.** no; **8.** 180° on Fahrenheit thermometer and 100° on Celsius thermometer; **9.** −40°; **10.** 100°F

Page 21. **B. 1.** no; **2.** yes; **3.** 37°C; **4.** no; **5.** no; **6.** no; **7.** 3; **8.** 50°C; **9.** 64°; **10.** 37°C; **11.** yes; **12.** 0°C; **13.** no; **14.** heating

Page 22. **A. 1.** a; **2.** b; **3.** b; **4.** b; **5.** a; **6.** a; **7.** a; **8.** b; **9.** a; **10.** a; **11.** a; **12.** a

Page 23. **B. 1.** 1 centimeter; **2.** 50 centigrams; **3.** 1 kilometer; **4.** 2 kiloliters. **C. 1.** 100; **2.** gram;

3. dag, hg, kg; **4.** 750; **5.** no; **6.** liter; **7.** 100 degrees; **8.** 37°C; **9.** hot; **10.** no

Page 25. **1.** 6.25; **2.** 27.36; **3.** 32.01; **4.** 4.2; **5.** 45; **6.** 15; **7.** 448; **8.** 1.8; **9.** 3.285; **10.** 19.25; **11.** .48; **12.** 15.2; **13.** .72; **14.** 16.48; **15.** 8.1; **16.** 130

Page 26. **1.** 356 inches; **2.** 39.6 meters; **3.** 34 centimeters; **4.** 29 yards; **5.** 104 m; **6.** 8 yd.; **7.** 6 ft.; **8.** 10.4 km; **9.** 4,000,000 square feet; **10.** 29.81 square meters; **11.** 602,176 square inches; **12.** .16 square centimeters; **13.** 2.25 square inches; **14.** 256 square yards; **15.** 1,303.21 square centimeters; **16.** 529 square kilometers;

Page 27. **17.** 15.36 square meters; **18.** 319.74 square feet; **19.** 2,258.16 square millimeters; **20.** 384 square inches; **21.** 96 square inches; **22.** 40.56 square centimeters; **23.** 9,126 square meters; **24.** 71.42 square millimeters; **25.** 1,331 cubic feet; **26.** 91.125 cubic centimeters; **27.** 9,528.128 cubic inches; **28.** 35.937 cubic meters; **29.** 24,389 cubic centimeters; **30.** 512 cubic inches; **31.** 1,728 cubic feet; **32.** 1.728 cubic meters; **33.** 1,000 cubic inches; **34.** 144 square feet

Page 28. **A.** **1.** 144 cm; **2.** 320 ft.;

Page 29. **3.** 14 in. × 14 in.; **4.** $1,144; **5.** 3,240 ft.; **6.** 192 ft.; **7.** 48 rolls; **8.** 10 km × 10 km; **9.** 4 ft. × 4 ft., 12 ft. × 12 ft., 20 ft. × 20 ft.; **10.** 20 in.; **11.** $2\frac{2}{3}$ in.; **12.** The perimeter doubles.

Page 30. **B.** **1.** 82 ft.; **2.** 360 m; **3.** 1,320 posts; **4.** 32 ft.; **5.** 9 ft.; **6.** 36 m; **7.** $24.96; **8.** 4.8 m; **9.** 33 meters; **10.** 336 feet

Page 31. **A.** **1.** 9 square feet; **2.** 144 square centimeters; **3.** 12 square feet (four times more area); **4.** 16 pints; **5.** 121 square feet ($13\frac{4}{9}$ square yards);

Page 32. **6.** 44.44 square yards ($44\frac{4}{9}$); **7.** 36 square inches; **8.** 32 square inches. **B.** **1.** 50 pounds; **2.** $63.20; **3.** 484 square feet; **4.** 64 square feet; **5.** 1,024; **6.** 490,000 square feet

Page 33. **A.** **1.** 6 square feet; **2.** 150 square yards; **3.** 253.5 square feet (36,504 square inches); **4.** 319.74 square inches; **5.** 486 square inches; **6.** 13.5 square meters; **7.** 15.36 square inches; **8.** .375 square centimeters; **9.** 864 square feet; **10.** 45.375 square feet; **11.** 600 square meters; **12.** 54 square inches; **13.** 1,278.96 square millimeters

Page 34. **B.** **1.** 1,350 square yards; **2.** 384 square centimeters; **3.** 6 square inches; **4.** $32\frac{2}{3}$ square feet (4,704 square inches); **5.** $66\frac{2}{3}$ square yards (600 square feet); **6.** $10\frac{2}{3}$ square yards (96 square feet); **7.** $37\frac{1}{2}$ square feet (5,400 square inches); **8.** 3,456 square feet; **9.** 1,350 square inches; **10.** 983.04 square centimeters; **11.** 96 square feet; **12.** 24 square meters; **13.** 294 square inches; **14.** 223.26 square centimeters; **15.** 96 square inches; **16.** 96 square meters; **17.** $13\frac{1}{2}$ square feet (1,944 square inches); **18.** 6 square meters; **19.** 61.44 square inches; **20.** 433.5 square millimeters; **21.** 864 square centimeters; **22.** 1,350 square feet; **23.** 24 square yards; **24.** 150 square inches; **25.** 294 square yards; **26.** 216 square feet; **27.** 34.56 square meters

Page 35. **A.** **1.** 3,276.8 bu.; **2.** 1,000 cubic yards; **3.** 4-inch cube; **4.** 480 gallons; **5.** $274\frac{5}{8}$ cubic feet (274.625); **6.** 512 cubic inches; **7.** 4,096 cubic centimeters; **8.** 27 cubic feet; **9.** 250.047 cubic centimeters

Page 36. **B.** **1.** 857.375 cubic feet; **2.** 3,375 cubic meters; **3.** $166\frac{3}{8}$ cubic feet; **4.** 729 cubic inches; **5.** 12,167 cubic feet; **6.** $12\frac{19}{27}$ cubic feet (21,952 cubic inches); **7.** 148,877 cubic inches; **8.** $37\frac{1}{27}$ cubic yards (1,000 cubic feet); **9.** 343 cubic centimeters; **10.** 21,024.576 cubic feet; **11.** 64 cubic inches; **12.** 1,000 cubic centimeters; **13.** 1 cubic foot; **14.** $42\frac{7}{8}$ cubic feet; **15.** 64 cubic meters; **16.** 8,000 cubic feet; **17.** 29,791 cubic centimeters; **18.** $91\frac{1}{8}$ cubic centimeters; **19.** $4\frac{17}{27}$ cubic feet (8,000 cubic inches); **20.** 343 cubic millimeters; **21.** 64 cubic feet; **22.** 8 cubic inches; **23.** 125 cubic centimeters; **24.** 216 cubic yards; **25.** 8 cubic feet; **26.** 3.375 cubic inches; **27.** 74.088 cubic centimeters

Page 37. **A.** **1.** 425 feet 4 inches; **2.** 384 inches; **3.** $33\frac{1}{3}$ yards; **4.** 10.8 meters; **5.** 13.6 centimeters. **B.** **1.** 256 square inches; **2.** 1,742,400 square feet; **3.** 761.76 square meters; **4.** 2.0449 square dekameters; **5.** 6.76 square meters; **6.** 289 square feet. **C.** **1.** 216 square feet; **2.** 96 square inches; **3.** 20.0934 square centimeters; **4.** 5,364.06 square meters; **5.** 1,014 square inches. **D.** **1.** 17.576 cubic inches; **2.** 2,197 cubic feet; **3.** 107,850.176 cubic centimeters; **4.** 157.464 cubic meters; **5.** 238.328 cubic meters

Page 38. **E.** **1.** 84 feet; **2.** 224 feet; **3.** 12 inches; **4.** 16 feet; **5.** 2,704 square inches; **6.** 7,056 square inches; **7.** 72 square inches; **8.** $1,078; **9.** 1,225 square feet; **10.** 125 square centimeters

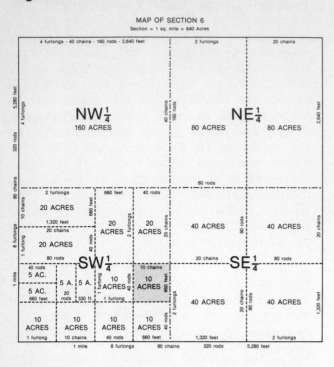

MAP OF SECTION 6
Section = 1 sq. mile = 640 Acres

The perimeter of the shaded parcel is 2,640 feet.

Page 40. **1.** 52 feet; **2.** 71 inches; **3.** 134 yards; **4.** 121 meters; **5.** 170 cm; **6.** 1,286 m; **7.** 186 in.; **8.** 234 ft. 6 inches; **9.** $93\frac{1}{2}$ square feet; **10.** 9.6 square meters; **11.** 6.24 square centimeters; **12.** 154 square inches; **13.** 5.2 square feet; **14.** 29.11 square inches; **15.** 253 square meters; **16.** 7.35 square centimeters;

Page 41. **17.** 52.5 cubic inches; **18.** 4.6875 cubic feet; **19.** 12 cubic meters; **20.** 10,584 cubic centimeters; **21.** 576 cubic meters; **22.** 18 cubic inches; **23.** 20.46 cubic centimeters; **24.** 22,496 cubic feet; **25.** a; **26.** b; **27.** c; **28.** b; **29.** d

Page 42. **A.** **1.** 920 feet; **2.** 36 inches; **3.** $10.35; **4.** $1; **5.** $1,353; **6.** 210 feet

Page 43. **B.** **1.** 480 ft.; **2.** 1,400 m; **3.** 20 in.; **4.** 24 km; **5.** 35 ft.; **6.** 42 in.; **7.** 22 m; **8.** 22 cm; **9.** 840 yd.; **10.** 40 ft.; **11.** 62 in.; **12.** 98 m

Page 44. **A.** **1.** $18\frac{2}{3}$ yards; **2.** $14.85; **3.** The 9 × 11 room is 3 square feet larger; **4.** 6,000 square feet; **5.** 6; **6.** 119 square inches; **7.** $56\frac{1}{4}$ pounds;

Page 45. **8.** 96 square feet. **B.** **1.** 4 pints; **2.** 120 square feet; **3.** 64 square inches; **4.** 1,800 square feet; **5.** $48; **6.** 104.5 square feet; **7.** $33\frac{1}{3}$ square yards; **8.** 71.06 square inches

Page 46. **A.** **1.** 40 loads; **2.** 12,600 gallons; **3.**

30 cubic feet; **4.** 6.75 cubic feet ($6\frac{3}{4}$); **5.** 2,400 boxes; **6.** 5 cubic feet (8,640 cubic inches); **7.** 12 cubic meters;

Page 47. **8.** $11\frac{1}{4}$ cubic feet; **9.** 9 cubic feet; **10.** 480 cubic inches. **B.** **1.** 6,336 cubic feet; **2.** 625 pounds; **3.** 16,800 cubic meters; **4.** 120 cubic centimeters; **5.** 640 cubic feet; **6.** $3,207.12

Page 48. **A.** **1.** 750 feet; **2.** 106 meters; **3.** 48 yards; **4.** 90 inches; **5.** 24 meters; **6.** 138 feet; **7.** 30 centimeters; **8.** 62 millimeters; **9.** 40 inches. **B.** **1.** 432 square centimeters; **2.** 383.5 square inches; **3.** 432 square feet;

Page 49. **4.** 16 square feet; **5.** 21 square inches; **6.** 108 square meters; **7.** 253 square feet; **8.** 243 square centimeters; **9.** 60 square yards. **C.** **1.** 64 cubic feet; **2.** 2,700 cubic inches; **3.** 640 cubic centimeters; **4.** 24 cubic inches; **5.** 40 cubic meters; **6.** 48 cubic feet

Page 50. **D.** **1.** a; **2.** b; **3.** d; **4.** a; **5.** b; **6.** a; **7.** c; **8.** c; **9.** b; **10.** c

Page 51. 10 sheets of paneling must be bought even though only $9\frac{1}{2}$ sheets will be used.

Page 52. **1.** obtuse; **2.** acute; **3.** straight; **4.** right; **5.** reflex; **6.** right; **7.** b; **8.** c; **9.** d; **10.** a; **11.** e; **12.** 27 square meters; **13.** 44 square feet;

Page 53. **14.** 15 square yards; **15.** 30 square inches; **16.** 33 feet; **17.** 41 meters; **18.** 32 inches; **19.** 22 yards; **20.** b; **21.** a; **22.** d

Page 54. **A.** **1.** right; **2.** acute; **3.** obtuse; **4.** obtuse; **5.** straight; **6.** reflex; **7.** acute; **8.** right; **9.** obtuse

Page 55. **B.** **1.** right; **2.** reflex; **3.** acute; **4.** straight; **5.** obtuse (Examples of these types of angles may vary as long as they meet the criterion for each type.)

Page 56. **A.** **1.** 14.72 km; **2.** 10 in.; **3.** 46 yd.; **4.** $15\frac{1}{2}$ ft.; **5.** $25\frac{1}{2}$ ft.; **6.** 26.5 m;

Page 57. **7.** $42\frac{1}{4}$ ft. (42 ft. 3 in.); **8.** 31 yd.; **9.** 49.5 cm; **10.** 215 ft.; **11.** 24 in.; **12.** 18 ft. **B.** **1.** 30.3 cm; **2.** 6.5 mi.; **3.** $2\frac{1}{8}$ in.; **4.** $492\frac{5}{6}$ feet (492 ft. 10 in.); **5.** $147\frac{1}{3}$ ft. (147 ft. 4 in.)

Page 58. **A.** **1.** 219 square feet; **2.** 10,127.52 square centimeters; **3.** 14 square miles; **4.** 416 square feet; **5.** $2,862;

Page 59. **6.** 3,984 square meters; **7.** 38,316 square feet; **8.** 896 square feet; **9.** 990 square inches ($6\frac{7}{8}$ square feet). **B. 1. a.** 45 square feet, **b.** 192 square inches;

Page 60. **c.** 1,200 square meters, **d.** 112 square feet; **2.** 600 square meters; **3.** 72 square feet; **4.** 27 square feet; **5.** 126 square feet; **6.** 8 square yards; **7.** 150 square feet.

Page 61. **A. 1. a.** less than 90°, **b.** exactly 90°, **c.** more than 90° but less than 180°, **d.** exactly 180°, **e.** more than 180° but less than 360°; **2.** p = a + b + c; **3. a.** 52 ft., **b.** 173 yd., **c.** 49 cm, **d.** 24 cm, **e.** 28 ft., **f.** 16 m; **4. a.** 630 square inches, **b.** 324 square yards, **c.** 450 square meters

Page 62. **B. 1.** e; **2.** c; **3.** a; **4.** b; **5.** d. **C. 1.** 12 square meters; **2.** 1.5 square feet (216 square inches); **3.** 162 square yards; **4.** 250 square feet. **D. 1.** 18 in.; **2.** 53 ft.

Page 63. 54 square yards

Page 64. **1.** $18\frac{6}{7}$ cm; **2.** $15\frac{5}{7}$ in.; **3.** 66 m; **4.** $12\frac{4}{7}$ mm; **5.** $50\frac{2}{7}$ in.; **6.** $163\frac{3}{7}$ ft.; **7.** 314 square inches; **8.** 1,017.36 square kilometers; **9.** 153.86 square centimeters; **10.** 6,500.585 square yards.

Page 65. **11.** 3,077.2 cubic feet; **12.** 2,009.6 cubic yards; **13.** 67,852.26 cubic feet; **14.** 1,727 cubic inches; **15.** b; **16.** c; **17.** a; **18.** a

Page 67. **A. 1.** 352 feet; **2.** 154 meters; **3.** $1\frac{4}{7}$ miles; **4.** $15\frac{5}{7}$ feet; **5.** $37\frac{5}{7}$ feet; **6.** 132 inches; **7.** 4,554 feet, **8.** $47\frac{1}{7}$ feet;

Page 68. **9.** 240 trips; **10.** $50\frac{2}{7}$ feet; **11.** 88 feet. **B. 1.** 28, 176; **2.** 7, 22; **3.** 12, $37\frac{5}{7}$; **4.** 70, 440; **5.** 196, 1,232. **6.** $106\frac{6}{7}$ yd.; **7.** $9\frac{3}{7}$ in.; **8.** $84\frac{6}{7}$ mi.; **9.** $6\frac{2}{7}$ in.; **10.** 44 cm; **11.** $292\frac{2}{7}$ ft.

Page 69. **A. 1.** 84, 5,544; **2.** 7, 154; **3.** $18\frac{2}{3}$, $273\frac{7}{9}$; **4.** $10\frac{1}{2}$, $86\frac{5}{8}$; **5.** 14, 616;

Page 70. **6.** 616 square inches; **7.** $50\frac{2}{7}$ square feet; **8.** $314\frac{2}{7}$ square inches; **9.** 154 square centimeters; **10.** $27,182\frac{4}{7}$ square yards; **11.** 1,386 square kilometers; **12.** 9,856 square inches; **13.** 3,850 square meters. **B. 1.** 18, 254.34; **2.** 200, 31,400; **3.** 5, $78\frac{4}{7}$; **4.** $10\frac{1}{2}$, $346\frac{1}{2}$; **5.** 30, 706.5;

Page 71. **6.** 154 square meters; **7.** $12\frac{4}{7}$ square inches; **8.** $28\frac{2}{7}$ square feet; **9.** 2,464 square centimeters; **10.** 2,464 square centimeters; **11.** $804\frac{4}{7}$ square inches; **12.** 1,386 square inches; **13.** 7,546

square centimeters; **14.** 12,474 square feet; **15.** 616 square millimeters

Page 72. **A. 1.** 3,465 gallons; **2.** 1,540 cubic meters; **3.** 22 cubic inches; **4.** 2,512 cubic yards;

Page 73. **5.** 9,240 gallons; **6.** 3,696 cubic inches; **7.** $1,178\frac{4}{7}$ cubic inches; **8.** $157\frac{1}{7}$ cubic feet; **9.** no. **B. 1.** $212\frac{1}{7}$ gallons; **2.** 100.48 cubic inches; **3.** 175.84 cubic inches; **4.** The capacity is $\frac{1}{2}$ as much (87.92 cubic inches);

Page 74. **5.** $410\frac{2}{3}$ cubic feet; **6.** 1,540 cubic yards; **7.** 10, $1,964\frac{2}{7}$; **8.** 8, $402\frac{2}{7}$; **9.** $10\frac{1}{2}$, $1,364\frac{11}{32}$; **10.** 9, $4,582\frac{2}{7}$; **11.** 28, 303,072; **12.** 42, 4,158; **13.** 14, 1,848; **14.** 20, $6,285\frac{5}{7}$; **15.** 9, $572\frac{11}{14}$; **16.** 12, $1,697\frac{1}{7}$; **17.** $4\frac{1}{2}$, $700\frac{1}{14}$; **18.** $3\frac{1}{7}$, 462; **19.** 4, $502\frac{6}{7}$; **20.** 10, $9,428\frac{6}{7}$; **21.** 9, $3,309\frac{3}{7}$; **22.** 24, 3,168; **23.** 17, $18,165\frac{5}{7}$; **24.** 8, $3,218\frac{2}{7}$; **25.** 18, $5600\frac{4}{7}$

Page 75. **A. 1.** circumference = 44 centimeters, area = 154 square centimeters; **2.** circumference = $6\frac{2}{7}$ inches, area = $3\frac{1}{7}$ square inches; **3.** circumference = $50\frac{2}{7}$ feet, area = $201\frac{1}{7}$ square feet; **4.** circumference = $18\frac{6}{7}$ yards, area = $28\frac{2}{7}$ square yards; **5.** circumference = 88 centimeters, area = 616 square centimeters; **6.** circumference = 176 inches, area = 2,464 square inches; **7.** circumference = 154 feet, area = $1,886\frac{1}{2}$ square feet; **8.** circumference = 132 meters, area = 1,386 square meters; **9.** circumference = 264 millimeters, area = 5,544 square millimeters; **10.** circumference = 132 feet, area = 1,386 square feet;

Page 76. **11.** circumference = 44 inches, area = 154 square inches; **12.** circumference = 176 centimeters, area = 2,464 square centimeters. **B. 1.** 924 cubic feet; **2.** $603\frac{3}{7}$ cubic yards; **3.** 7,392 cubic meters; **4.** $100\frac{4}{7}$ cubic inches; **5.** 5,390 cubic feet; **6.** 16,632 cubic inches; **7.** 12,320 cubic centimeters; **8.** 198 cubic meters; **9.** 21,560 cubic inches

Page 77. **C. 1.** b; **2.** c; **3.** a; **4.** a; **5.** d (The numerical result of your computations should be $1\frac{185}{189}$ cubic yards.); **6. a.** $1,018\frac{2}{7}$ square inches, **b.** 154 square centimeters, **c.** 154 square centimeters, **d.** $38\frac{1}{2}$ square feet, **e.** 616 square meters, **f.** 1,386 square centimeters;

Page 78. **7. a.** 66 meters, **b.** 44 feet, **c.** 88 inches, **d.** 22 meters, **e.** 88 centimeters, **f.** 308 feet; **8. a.** $37\frac{5}{7}$ cubic inches, **b.** $402\frac{2}{7}$ cubic feet, **c.** 396 cubic centimeters, **d.** 550 cubic inches

Page 80. **1.** b; **2.** b; **3.** a; **4.** c; **5.** a; **6.** a; **7.**

a; **8.** a; **9.** c; **10.** c; **11.** b; **12.** a; **13.** c; **14.** b;

Page 81. **15.** a; **16.** a; **17.** b; **18.** a; **19.** b; **20.** b; **21.** c; **22.** c; **23.** a; **24.** b; **25.** c;

Page 82. **26.** c; **27.** a; **28.** c; **29.** a; **30.** b; **31.** c; **32.** a;

Page 83. **33.** b; **34.** a; **35.** c; **36.** c; **37.** b; **38.** b; **39.** a; **40.** c;

Page 84. **41.** a; **42.** c; **43.** a; **44.** b; **45.** c; **46.** b;

Page 85. **47.** a; **48.** c; **49.** c; **50.** b; **51.** a;

Page 86. **52.** c; **53.** c; **54.** b; **55.** c; **56.** a; **57.** a;

Page 87. **58.** c; **59.** a; **60.** b; **61.** d; **62.** no; **63.** b; **64.** a